The Doctors' Plague

GENERAL EDITORS: EDWIN BARBER AND JESSE COHEN

GREAT DISCOVERIES

SHERWIN B. NULAND

The Doctors' Plague

Germs, Childbed Fever, and the Strange Story of Ignác Semmelweis

ATLAS BOOKS

W. W. NORTON & COMPANY
NEW YORK · LONDON

For information about permission to reproduce selections
from this book, write to Permissions, W. W. Norton & Company, Inc.,
500 Fifth Avenue, New York, NY 10110

Manufacturing by the Haddon Craftsmen, Inc.
Book design by Chris Welch
Production manager: Julia Druskin

Library of Congress Cataloging-in-Publication Data

Nuland, Sherwin B.
The doctors' plague : germs, childbed fever, and the strange story of
Ignac Semmelweis / Sherwin B. Nuland.—1st ed.
p. cm.—(Great discoveries)
Includes bibliographical references.
ISBN 0-393-05299-0 (hardcover)
1. Semmelweis, Ignâac Fèulèop, 1818-1865. 2. Puerperal septicemia.
3. Asepsis and antisepsis. I. Title. II. Series.
RG811.N85 2003
618.7'4'0092—dc21

2003011412

W. W. Norton & Company, Inc., 500 Fifth Avenue, New York, N.Y. 10110
www.wwnorton.com

W. W. Norton & Company Ltd., Castle House, 75/76 Wells Street, London W1T 3QT

1 2 3 4 5 6 7 8 9 0

For Ferenc Gyorgyey
Our great friendship began with Semmelweis

There is a "word of fear" that I shall pronounce when I utter the name of Puerperal Fever; for there is almost no acute disease that is more terrible than this. . . . There is something so touching in the death of a woman who has recently given birth to her child; something so mournful in the disappointment of cherished hopes; something so pitiful in the deserted condition of the new-born helpless creature, for ever deprived of those tender cares and caresses that are necessary for it—that the hardest heart is sensible to the catastrophe. It is a sort of desecration.

—Charles Delucena Meigs, Professor of Midwifery and the Diseases of Women and Children, Jefferson Medical College, 1851

The Doctors' Plague

I

She felt lucky in at least this one thing—it was a Sunday, and she would not have to go to the hospital alone. The Allgemeine Krankenhaus, that great forbidding place of multiple and enclosed buildings, walkways, and inner courtyards, had for months been the stuff of her worst fears. But now that the time had come to surrender herself to it, the girl felt strangely relieved, with the sense that perhaps fortune was again beginning to smile on her just a little. Sunday meant that her friend Liesl was at home, freed for a day from the large, airless loft where the head seamstress, Frau Eigenbrodt, supervised some twenty girls at one of the most fashionable dressmaking houses in mid-nineteenth-century Vienna. Liesl would accompany her as far as the hospital's Eighth Courtyard, help her up the few outer steps, and knock on the large wooden door to the lying-in unit. With such a good friend to take her that far, the fear would be lessened. She would part from Liesl when the nurses appeared, and not lose courage to face the birth of her child.

The girl had stood in the courtyard alone and unobserved many times in these past weeks, watching other young women make their difficult way up the stone outer steps in the final days of their pregnancies. She had even smiled now and then in the midst of her unhappiness, upon witnessing the kindness on the faces greeting these girls and women heavy with the bulk of an infant soon to be born. She knew that the nurses would treat her kindly too, once she crossed the threshold and committed herself to their care.

Kindness was what the girl needed at that long-awaited, long-dreaded hour. Kindness had been all but a stranger to her life since she left her father's house, five months before. The happiness of being the adored only child of a prosperous, warmhearted tradesman had in an instant evaporated, when she dared to reveal that she was with child. She had somehow expected something else from him, something more of understanding and comfort. Everything that had passed between father and daughter in her eighteen years led her to anticipate a different kind of response, even a soothing of her anguished guilt. Papa would know what to say, she had thought. He would know how to speak to her in the gentle tones that so many times in the past had brought consolation in hours of sadness and even grief, as when her beloved mother died of a lung infection when the girl was only twelve. And he would also know what should be done to make things better. Perhaps, she had hoped, he would find some means of making this thing go away, making it disappear as though it had never been. But there had instead been that outburst of unanticipated rage. There was no mother to calm the storm of abuse and insistent questioning hurled at her by the infuriated widower—only the servant girl Maria, as uncomprehending and stunned by the force of the tirade as she was herself.

Everything was made worse when she refused to give the father's name. What good would it have done? He was a student of philosophy at the University of Vienna, and she had met him in a coffeehouse one afternoon quite by accident, after one of her singing lessons. It was thrilling in those first weeks, to go off with such a shining and silver-tongued youth into the meadows outside the city. They would speak of poetry and the romance-scented trivia that pass so effortlessly between young people when they are first finding a way into each other's hearts. And then early one evening when her Papa thought she was at a friend's house, she gave herself to the ardent boy, right there in the fields under a setting sun. They had been speaking of Goethe, and the young philosopher was comparing his lovesickness to Werther's.

She at first had no regrets for her impulsive behavior, only a sense of wonder and a flowering joy whose petals were opening toward even better days to come. But the boy was inexplicably hesitant to go with her to the pretty little house on the Gartenstrasse to meet Papa. And then, in too short a time, everything began to change between them. Within only a few weeks, he seemed to have undergone a transformation. To make love was all he now wanted to do, and he found rooming houses and the garrets of friends where they met furtively and only long enough to rut, because rutting is what it soon became. She did not know how to stop herself or him, and before long her burgeoning love for the student had become shame. When she told him she was pregnant, he turned from her in disgust, saying it was his own fault for dallying with such a stupid girl—"dallying" was the word he used, and with that word told her what she had begun to suspect.

And so she went to her father, truly believing that he would put his protecting arms around her as he had always done since she was a honey-haired little girl, and reassure her that all would be well. But she was stupefied by his outrage, and even now, as she sat uncomfortably on the side of her narrow cot in Liesl's tiny room, feeling the earliest of her waves of pain, she could not bring herself to believe that her beloved Papa had actually ordered her out of his house, bellowing his rage in a torrent of abuse. And she could not believe that he had refused to acknowledge the tear-stained letters she had written him in the following weeks, begging to be allowed to come home, begging for solace. Now her baby would have to be born in secret, in the anonymity and bustle of a large public hospital where no one knew her, and she knew no one.

But the girl cherished one glimmering iota of hope—maybe, just maybe, she would go to Papa after the babe was born and he would forgive her everything, once he had laid eyes on his grandchild and loved it even as he would love it if she were properly married.

Liesl was the daughter of the housekeeper who had been with the family until a year after Mama died. The two girls had been friends since earliest childhood and remained close even after Liesl's mother left to work in one of the new factories then being built in Vienna. And it was in Liesl's little chamber under the eaves of a rooming house for working girls that she had been living since that awful day when her life changed.

Deep in thought, she sat on the edge of her bed for about fifteen minutes; its sheets were soaked with her waters. When

Allgemeine Krankenhaus (General Hospital), Vienna, 1825. (©
Austrian Archives/Corbis)

the next pain began, she leaned over to touch her sleeping
friend's arm. Without a word, Liesl got up, wrapped her thin
blanket around the girl's shoulders and then dressed herself
in a threadbare robe draped over her nightdress.Wordlessly,
the two young women made their way down the four long
flights of battered wooden stairs and out into the dawning
day's light. Liesl turned to the girl as they stood for a moment
in front of the house, gave her a wide smile of encouragement, and took her elbow, in preparation for the long walk
of more than half a mile to the Allgemeine Krankenhaus, the
General Hospital of Vienna.

It mattered to neither of them on that late-May morning
that they were barefoot as they trod cautiously on the litter-
strewn pavement. It took them nearly an hour to reach their
destination. They were scarcely noticed by the attendant at

the sprawling institution's front lodge, accustomed as he was to the sight of bulging women being brought past his station on their way to the lying-in unit. By the time they reached the Eighth Courtyard, the girl was exhausted. She had almost welcomed the onset of each intermittent contraction within her, because it provided an opportunity to stop for a few moments' rest. But the pains were becoming stronger now, and more frequent. It was only with difficulty that Liesl got her friend up the steps to the huge oaken door.

A kindly-faced nurse appeared, and then another. Perhaps it was the clothing they wore: the soft bulk of their long uniform dresses of dark blue flannel, billowed out below the waist by underlying petticoats; the starched whiteness of the wide apron that covered each of them from breast to ankle; the fluffy linen caps, tied under their chins by cloth strings. Whatever it was, these two women looked protective and even motherly. Taking the girl by each arm, they led her slowly up the long staircase at the very top of which was a small table where a skimpily bearded medical student sat perusing a large notebook spread out before him, through thick, wire-framed spectacles. The girl suddenly realized that she had been so distracted by seeing the two nurses that she had not even said good-bye to Liesl. No matter. Her friend would understand, and they could laugh about it a few days later, when Liesl would visit and see the new baby for the first time.

When the three women reached the landing, the student looked up from the scrawled pages and pointed to his right. "Take her to the First Division, please," he called out loudly in a marked Moravian accent, and the girl hesitated because she had heard that she must specifically ask to be admitted to the unit where the deliveries were done by midwives. In the other of the two lying-in divisions, the students and supervising physicians delivered the babies, which meant many more examinations during the course of labor and therefore many exploring fingers intruding into the birth canal. The student midwives and their teachers seemed content with far fewer interventions, and this information was well known among Viennese women. Liesl had heard it spoken of by the seamstresses at Frau Eigenbrodt's, and told her friend to be sure that she did not find herself in the prying hands of the medical students, or rather that their prying hands not find their way into her. Mothers who had already borne several children seemed not to care, said Liesl, "but this is not the way it should be for a young woman like you. Make sure your child is brought into this world by someone who won't poke into you every time he gets curious about how things are going, or just because he wants to learn something."

"Please tell me, Nurse," the girl asked. "Is the First Division for the doctors or the nurses?" The reply was maternal not only in its soft tone but also in the unyielding sound of it. "That shouldn't concern you, my dear," said the older of her attendants, "but if you must know, it's the one where the doctors are in charge."

"But please, madam, let me go to the other ward. I want my baby to be delivered by a midwife."

"I'm sorry, young woman, that won't be possible. The hospital has rules, and you must follow them."

At this point, the medical student felt called upon to exhibit his authority.

"Please pay attention, young miss! Patients are divided equally and alternately by the days on which they come to us.

From Friday afternoon until Sunday afternoon, we admit to the First Division. You must go to the doctors!"

She tried to plead with him.

"But please, good sir, can't it be that I . . . ?"

"No, it cannot. It absolutely *cannot*! Now be a good little woman and go with the sister."

She burst into tears, and tried to pull herself from the arm of the older of the two nurses, which the woman had placed lightly around her shoulders as though to comfort and restrain her at the same time, while exerting just enough pressure to turn her in the direction of the First Division. But at this very moment the girl was distracted by the powerful wave of a contraction and could think of nothing else. When it had subsided, she found herself being guided toward a short corridor beyond which she could see the ward, a large, open, rectangular space surrounded on all four sides by beds placed vertical to the whitewashed walls, in each of which lay a big-bellied woman. Some of the women were in labor. Medical students and doctors seemed to be everywhere. There must have been a dozen of them. Some were standing by themselves and writing notes, while a few others stood conversing quietly in a small group alongside one of the fireplaces at each end of the ward. At each of three beds, a student was examining a woman whose bent legs were being drawn backward by a nurse on either side. Near one of them stood a man in his thirties, who appeared to be instructing his young charge.

"You mustn't worry, dear," said the nurse who was leading the girl toward the only empty bed in the ward. "Things are not always this busy. It's early, and the students have all just arrived from the deadhouse, so they're keen to do some internal examinations. They go there first thing every morning, and then they come up here as eager as puppies. Soon most of them will be gone, because it's Sunday."

The girl knew nothing of deadhouses, and the thought terrified her. As the nurse was getting her settled onto the rumpled sheet covering the thinly stuffed mattress of the bed, she asked apprehensively what such a horrid-sounding thing was.

"Doctors must learn from cutting open the dead, you know, so they do that every morning before beginning their regular work up here."

"But who dies? Please don't tell me I could die having a baby. Oh, please, please, Nurse, I couldn't bear to think of it."

The nurse's serene expression clouded over, but just for a moment. "Hush, hush," she said. "You'll get yourself all worried and excited for no reason. Surely you know that every once in a great while a mother dies giving birth. But this is a great and famous hospital, the finest in all Europe. So you mustn't concern yourself with such things as death. After all, these are modern times—it is 1847. Be calm, my child. We'll take good care of you."

Reassured, the girl did as she was bidden. She was tired from her long trek to the hospital, and it felt good to be in bed at last. When the nurse had finished asking a series of admissions questions, she closed her eyes, hoping to get a few minutes' sleep before the next pain. But her tranquillity did not last long. Within minutes, a student appeared at the foot of her bed, clearing his throat to awaken her. Without apologizing for the intrusion, he called for nurses to assist him in the first examination of his new patient.

It was obvious that he was new at this. Before he had even gotten under way, the instructor came hurrying over and whispered a few words in his ear. The examination seemed interminable, but at first the embarrassment was worse than the discomfort. The instructor kept making the student stop and then begin again with a slightly different way of probing. Sometimes he would demonstrate a move himself, inserting his long, spidery fingers far into the girl and pushing on something deeply within. Each time, it made her feel like screaming with the strangeness of the sensation and the fear that something dire was about to happen. The student's clumsy hands were like spades, and when he pushed his blunt, bare fingers toward their uncertain destination, the girl was sure that he must be intent on digging something out of her. She was not calmed by hearing his teacher whisper harshly, "Gentle, gentle, you blockhead. The idea is to do no harm. Remember Hippocrates, dolt." But to the girl he said only, "You must stop squirming, young woman, or we cannot do what is necessary. It is for your own good." Beyond those few words she might as well not have existed in the minds of those two men, but for her organs of reproduction. She found it impossible to lie back as she was told, and kept trying to get up. Each time she strained forward at the waist, one of the nurses would push her gently but firmly downward onto the pillow. Only by clenching her jaws was she able to keep from crying out. But she could not suppress a deep-throated groan whenever one of the men thrust his searching fingers forcefully against the opening of her womb.

When the examination was finally over, the girl lay back exhausted. Her tormentors stood at the foot of the bed discussing their findings as though she were not there. She could barely hear them and, in any event, was by then past caring about anything except having this birth over with. But she was only at the very beginning of her travail. Nodding toward her, the student asked something of his instructor, to which assent was given. One of the nurses seemed to disagree with his decision and remonstrated with the older man, but it did no good. "Can't it wait? The poor thing has just gotten here, and she needs some time to rest and get used to us."

It was to no avail. The student beckoned to one of his young colleagues on the other side of the ward, who came over without delay. And then, unbelievable as it was to the girl, he began demonstrating something within his new patient to the friend, which he had just learned while deep in her body. The other fellow now just had immediately to try it himself, which he proceeded to do as the instructor smiled approvingly. The nurses stood there in stony-faced, infuriated silence.

At last, the girl was left alone. She felt as though beaten, bruised inside and out, and probably oozing blood from lacerated surfaces she had never seen. Sobbing softly, she whispered as if to her father, making believe he sat with her at the bedside holding her hand. "Oh, my darling, darling Papa, what have I done to you, what have I done to myself? Please, please, you must come to me, help me, take me away. Take me out of this wretched place, or I will die. Oh, my Papa. I love you so—if only I could see your beloved face for a single moment, all of this would be so much easier. I cannot bear this without you to help me."

She tried to comfort herself by remembering the purpose of it all. In only a short time it would be over, and she would be holding in her arms a beautiful baby to bring purpose and joy to her life, and return her to the home illumined by love. The life that had once been hers would be hers again—and with the beaming happiness of Papa to share it. Lying on the rumpled sheets of the narrow hospital bed, now saturated with her sweat and soiled down below by blood and unseen discharges, she imagined herself in her own soft bed at home, but now cuddling an infant in her arms. And she fell asleep until the next pain woke her up a few minutes later.

The labor lasted fourteen hours. Again and again, the student's hands entered the girl's body, and several times his teacher's did the same. Finally, at about 10:00 P.M., a new instructor determined that it was time to go to the small room off the ward where the babies were born. The birth was normal in every way, and the newborn boy was as beautiful in his mother's eyes as she had expected he would be. She could not wait until Tuesday evening, the earliest visiting hours when Liesl might come after her long day at Frau Eigenbrodt's, to see the baby for the first time. And then, perhaps a few days afterward, her friend's mother could be there, and together they would discuss the best way to plan a return home to Papa. It was a rosy time, and the future was beginning to look like the past had once been.

The girl awoke very early the next morning. She was in another large ward, very much like the one to which she had at first been admitted. Battered and torn as she felt in that underplace of wondrous birth within her, her thoughts were only of the tiny boy who would soon be brought to her so that she could hold him in her arms. He was to be christened Ferdinand, because that was Papa's name, and the emperor's too. It had a manly sound, and he would grow up to be like his grandfather, strong and gentle, as a man should be. How pleased Papa would look when he first saw the child, and how proud that such a handsome boy should bear his name.

She had hoped for kindness from the nurses, and they fulfilled her every expectation. Patiently and with great concern for her comfort, one of the older sisters showed her how to hold the babe to her breast, and then instructed her in the feeding methods she would need to know within a day or two when her milk came in. She was exhilarated—intoxicated with the joy of motherhood and the realization of her child's complete dependence on her. What she felt for him was a kind of love she had never known and was totally unprepared for. It made her a woman.

Little Ferdinand stayed with her for a brief, lovely while before he was taken away again. Everything about being with the child felt so natural, as though she had been born for it. She was too preoccupied with her happiness to pay more than the slightest heed to the increasing discomfort in her lower abdomen. It was not until late in the afternoon that she thought to touch herself there, and was surprised at how the slightest pressure of her fingers worsened the pain. When her supper of thin soup and a boiled potato was brought, she was unable to eat it. The food remained untouched on the small table near the head of her bed. It was an hour before the orderlies came to remove it, and by that time she was nauseated at the sight of the potato and greasy soup. When she vomited in the late evening, it was surprising to her that bits of the morning's breakfast roll were in it.

The vomiting brought one of the nurses, who took her wrist and counted the pulse. She said that it was 100, and went off to fetch a doctor. Perhaps it was just unnecessary worry at the onset of these changes in her sense of well-being, but the girl was beginning to feel oddly cold, as though she was developing a chill. When the doctor arrived—a tall, long-legged man resembling nothing so much as a middle-aged giraffe dressed up in a soiled frock coat—he pulled down the sheet and stared for a long time at her belly. She thought she heard him say, "It's a little early for this to be happening," but it was not possible to be sure. He had the nurse separate her legs as he bent over to peer at the discharge. His head went lower, and she could tell that he was sniffing, obviously trying to detect an odor.

Was it a look of concern that she saw on his face when he straightened up? She turned her head from him to the nurse and back again, seeking some sign that might betray their thoughts. They were both trying to appear impassive, but the doctor's barely audible words gave him away. "We'll just have to watch carefully for the next few hours," he said to the nurse. "It's all right to bring the baby to her if she asks for it."

But when the child was brought an hour later, she could not find a comfortable way to hold him; every movement made her belly hurt. Seeming to sense the awkwardness, the baby became fitful. When he was taken away crying loudly, the girl tremulously asked the more experienced mothers on either side of her what might be happening, but their answers seemed evasive. She tried for more than an hour to calm herself, unsuccessfully. Her anxiety was mounting as the pain in her abdomen gradually worsened. An increasing thirst made her want to drink water, but she could only sip small amounts at a time. Her body was soon shivering with cold, making the glass shake in her trembling fingers. When she looked at her hands, she saw a blue color under the icy fingernails. It was at this point that she had a great shaking, teeth-chattering chill that made the bed rock beneath her. When it was over after about ten minutes, she became very hot and flushed with fever.

Bit by bit, she could feel her belly filling up with gas. Two more hours passed. Her skin was now wet with a clammy perspiration that attracted the attention of a nearby nurse, who took her pulse again. Once more, the doctor was called. This time when he lifted the sheet to examine her now distended abdomen, there was no need to sniff—she could smell a foul odor arising from between her legs. She cried out in startled pain a moment later when he pressed his probing hand into her bloated belly, and she knew that something was terribly wrong. Her pulse was now 120. The doctor called it "thin and small," letting her wrist drop from his hand, all the while staring with deep concern at her wan face as he distractedly slipped his watch back into the pocket of his waistcoat. When he looked at her tongue, she could tell that it was dry, whitened, and coated as though with a coarse fur. She had another shaking chill.

Everything was happening so quickly. She was becoming restless, taking short, rapid breaths. The doctor left her for an hour before returning to take her pulse again, this time feeling it for two full minutes. He gazed portentously at the small amount of dark, foul-smelling urine in the chamber pot underneath her bed. When he was through, he nodded meaningfully to the worried-looking nurse, and within a few minutes two orderlies had arrived with a stretcher, to which they rapidly transferred her. With the doctor and several nurses standing by, she was taken to a small anteroom off the main ward, but by this time she was too removed from clarity of thought to ask why. She tried to conjure up the face of her Papa, but it would not come to her; the baby seemed far away, a thing she had only imagined. When the doctor took a lancet from the small tray brought to him by a nurse and used it to

open a forearm vein for bleeding, she observed his move-
ments uncaringly, as though it were being done to someone
else. Listlessly, she watched her dark red blood spill into a
shallow metal bowl. The large, warmed turpentine poultices
placed on her increasingly bloated abdomen did nothing for
the pain, which would have been unbearable had she still
been alert to the full force of it. The next great chill had less
frightening an effect on her than the earlier ones, because her
perceptions were weakening. When the convulsive shaking
was over, she seemed to forget it had taken place. The nurses
tried to get her to swallow a mixture of calomel and opium,
but it merely made her vomit again.

Two of the nurses stayed at her bedside, speaking softly to
her but getting only incoherent responses now. She shouted
with the abdominal pain when they tried to turn her for an
enema, which they soon abandoned. Alternating between
dulled fearfulness and periods of passivity to the point of
barely being aware of her suroundings, she was vaguely conscious of a pounding headache. Two hours later, she barely
heard when the doctor announced her pulse to be 130, and
she was beyond caring. Her long blond hair was matted with
perspiration, soaking her pillow. The deep brown eyes
remained open, but glazed and recognizing nothing of their
surroundings. She was unaware of the next shaking chill. She
had become insensitive to everything.

For almost two days, the girl lay there in a sweat-drenched
coma that alternated with brief periods of delirium, during
which she clutched wildly at the sheets and shouted incomprehensible words to people who were not there. Except to
wipe the perspiration from her drawn face and clean her
body of its wastes, there was nothing the doctors or nurses

could do for her. She was beyond the help of human hands. When Liesl came to visit on Tuesday evening, she was turned away at the large entrance door where the two young women had last stood together on Sunday morning. She was told that her friend had asked that there be no visitors.

And then, shortly after twilight on Wednesday evening— three days after her baby's birth—the girl suddenly sat bolt upright in bed, her eyes staring wildly forward. With arms outstretched, she screamed "Papa, forgive me!" and fell back on the pillow, dead.

II

The physicians and nurses caring for the girl were all too familiar with the disease that took her life. In that year, 1847, one of every six mothers delivered in the First Division of the Allgemeine Krankenhaus was dying of it, and numerous others developed its earliest symptoms but somehow overcame them and avoided a similar fate. And the experience in Vienna was hardly unique; it was being duplicated in hospitals throughout Europe and, to a lesser extent, in America. Childbed fever was ubiquitous.

Every doctor at the medical school of the University of Vienna, of which the great hospital was a part, knew what he would see—and smell—upon opening the body of one of the many mothers who were brought to the autopsy tables of the deadhouse. Though there was a spectrum of variations, the underlying pattern was always the same. The incision into the abdomen would release a stench so foul that medical students were known to vomit and even faint on encountering it for the first time. It was the stink of putrid flesh and the thin, fetid fluid that emanates from it. And it was also the stink of whitish or discolored thick pus, of which there might be an abundance or only a small collection alongside the most affected of the organs. Always there would be an intensity of swollen inflammation of the uterus, extending into the tubes and ovaries and along the veins rising upward from the pelvis. The lining of the abdomen—the peritoneum—was thickened and shaggy, as were the ligaments, folds, and membranes that attach the female organs to one another and to the adjacent bladder and loops of bowel. The pus might lie freely in pools alongside one or another structure, or it might be pocketed into abscesses discoverable anywhere in the abdominal cavity. Moreover, the abdomen was not the only possible location for these lethal bags of purulent and repulsive fluid. They were sometimes hidden within the chest or the tissues of the body wall and even in the joints, because the foulness had entered the bloodstream and been disseminated like a spray of death.

The vagina and the external genitals often had the appearance of monstrously infected wounds, as though beaten with a club and allowed to fester into the discolored inner thighs. Commonly, the surrounding skin and underlying flesh of this area was so choked with foul fluid and gaseous bubbles that it crackled like a wad of crumpled newspaper under the pressure of an examining fingertip pressed into the skin. Should one turn away for a moment from these horrifying sights and let a glance fall on the face of the dead mother, no relief was to be found there. One saw a young woman who had become aged in a matter of days, the cheeks so recently flushed with the joyous anticipation of motherhood now sunken and sallow. This was the gray visage of suffering, the mask of death, the wages of childbed fever. No one knew what caused it, and no one knew what to do about it.

As to causes, there were, of course, many theories. One of the oldest—and to some physicians still the most persuasive—was suppression of the free discharge of lochia, the fluid that emanates from the uterus after a normal delivery. Long before the nature of inflammation or infection was understood, the appearance of the fluid and pus inside the abdomen of a woman dead of childbed fever was deemed to be so similar to lochia, that they were thought to originate in that source. According to this theory, lochia prevented from their free egress through the uterine outlet would stagnate, putrefy, and back up into the tissues and blood, causing pain, fever, and finally death. Some authorities believed that impurities of the blood accumulated during the nine months of pregnancy, perhaps because the enlarging uterus pressed on the intestines to cause stasis of fecal material, whose poisons were then absorbed into the veins. If the free flow of lochia did not purge the blood of these impurities after childbirth, the result was the general corruption called childbed, or puerperal, fever.

As to the reasons for the suppression, here, too, many theories were promulgated over the centuries. They were summed up by a seventeenth-century French physician who went by the Latinized name of Riverius: "The causes of this suppression are a too great thickness of the blood, a narrowness or obstruction of the vessels, cold air inadvertently received into the uterus, which closes the orifices of the vessels; taking cold at the feet, drinking of cold water; fear, terror, grief and other passions of the mind, which withdraw the course of the blood from the uterus," thereby preventing its poisons from being discharged into the lochia. This was tantamount to saying what so many believed, that virtually any shock to the system was capable of suppressing the lochia.

The notion of suppressed lochia had been the dominant theory of the classical period. The first known description of childbed fever is, not unexpectedly, to be found in the miscellany of texts known as the Hippocratic Corpus, long attributed to the Father of Medicine but now known to have had multiple contributors. The author of the Hippocratic volume called Book 1 of *Epidemics* describes the case of "Thasus, the wife of Philinus, having been delivered of a daughter, the lochial discharge being natural," who two weeks later was "seized with fever attended with rigor [shaking chills]," as well as pains in the abdomen and genital organs while her "lochial discharge ceased." Using what must have been the treatment of the time, the Hippocratic physician inserted a pessary, doubtless in the hope that supporting the uterus might permit proper drainage of the lochia. And indeed, "all these symptoms were alleviated." But only temporarily. The patient gradually worsened again, with the pain, fever, and rigors resuming and now accompanied by delirium. Reinsertion of the pessary did not help this time, and Thasus died on the twentieth day of symptoms after three days of coma. Cases like these were no doubt the basis on which the Hippocratic text *De Mulierum Morbis* described a typical case of childbed fever as it was seen some 2,200 years before the epidemics at the Allgemeine Krankenhaus. It is a testament to the storied observational powers of the Hippocratic physicians that their account so closely matches the descriptions of the cases seen in Vienna.

If, however, the purgation of the puerperium [the period after delivery] does not take place, the next thing is likely to be a fever with a chill, while the abdomen becomes swollen. If you touch the patient the whole body feels the pain of it, and especially if one should touch the abdomen. At the same time also a burning sensation in the abdomen is noted and there are apt to be pains in the loins. The patient's appetite becomes finicky, and wakefulness and irritability occur. . . . The urine is not unlike that of an ass. . . .

The pulse is always weak, at times also quick, more rarely strong, but thin. This occurs at the beginning of the disease and is apt to continue. After an interval the hollow parts of the face become red.

It was observations like these that led to the Forty-third Aphorism of Hippocrates: "If erysipelas [a term that at the time meant spreading inflammation] of the womb seize a woman with child, it will probably prove fatal."

Although the principle of lochial suppression remained the most widely accepted explanation for childbed fever, other theories were brought forth when the rudiments of science began to enter the thinking of physicians. The seventeenth century is often—and correctly—called the century of science or the era of the scientific revolution, but much of the theorizing of that intellectually exhilarating era was based on an uneasy mixture of meticulous observation confounded by unprovable speculation. Among the most prominent of the quasi-scientific formulations was the so-called milk-metastasis theory of childbed fever, said by its sponsors to be quite logical but in fact just as hypothetical as lochial suppression, and based on similar erroneous concepts.

It had long been believed that the milk of a nursing mother is composed of transformed menstrual fluid, which supposedly reaches the breast via a duct leading from the top of the uterus to the tip of the nipple. Even the usually clear-thinking Leonardo da Vinci believed in this hypothetical physiology, though his dissections had failed to demonstrate the existence of such a duct. Uncharacteristically, he was nevertheless so convinced that it must be present that he depicted it in an illustration destined to be among his most famous, the so-called coitus drawing.

When the abdomens of childbed fever victims were opened, the pus and the infected fluid that were found collected there appeared so similar to milk that some dissectors insisted that it had been rerouted away from its normal path to the breast. This seemed a logical way to explain why lactation slowed down and then stopped when a postpartum woman developed the disease: milk was being forced downward by some obstructive process, toward the pelvic cavity and its vessels. After entering the circulation, it would then be carried to various parts of the body. This was said to be the reason for the distant collections of foul fluid in the chest or joints discovered at autopsy. Like so many theories based on logic thrown awry by erroneous interpreting of observations, this one had a certain appeal to minds that had not yet developed the understanding that scientific beliefs must be based on evidence much less tenuous. The milk-metastasis theory received a huge boost in 1746 during the first documented hospital epidemic of childbed fever ever reported, at the Hôtel-Dieu in Paris. After dissecting many of the bodies made available by the carnage, three physicians published a report in which they described finding "a free lactescent fluid in the lower portion of the abdominal cavity and clotted milk adherent to the intestines." Of course, it was not clotted milk at all, but pus and whitish, infected fluid. Almost half a century passed before discerning observers pointed out its real nature.

The epidemic at the Hôtel-Dieu was the first of many that would plague hospitals for the following century and more. Though the institution was founded in 660, its main buildings were not constructed until the reign of Louis XIII, between 1610 and 1643. At the age of approximately one hundred when the epidemic took place, they were among the oldest of the large hospital facilities in Europe, of which there were still only a small number. But the situation was rapidly changing, and for young women susceptible to puerperal fever, this was not necessarily a good thing. More hospitals meant more epidemics.

Many new hospitals were built during the eighteenth century. Thirty-two were founded in the provinces of England between 1736 and 1799, as well as five in London from 1720 to 1745: the Westminster, Guy's, St. George's, the London, and the Middlesex. Much the same outburst of construction was taking place on the continent of Europe. The enormous Allgemeine Krankenhaus opened its doors in 1784.

The new hospitals were manifestations of an arising consciousness, on the part of governments and individuals, of society's responsibility to the poor. Spurred on at first by generalized social changes and then by the emerging industrial revolution, people were flocking to the cities, where they frequently found themselves in precarious circumstances. Crowded into areas meant to house far fewer, they were subject to living conditions that bred disease. Contagion and injuries were rampant, and the need for medical care seemed to grow with each passing year. Though the rich were cared for at home, that option was rarely available in the cramped and squalid circumstances in which the poor were forced to live. The older institutions, such as St. Bartholomew's in London and the Hôtel-Dieu in Paris, were no longer capable of carrying the load.

In England, associations of philanthropists banded together to establish hospitals, though some, like Guy's, were founded by individuals of great wealth. On the Continent, the new institutions were more likely to be sponsored by the state. Either way, they were becoming the locus for the teaching and learning so necessary to the changing ways of medical theory and practice.

With their large concentrations of the sick poor, hospitals were the ideal training grounds for the teaching of young men who wished to become physicians. This purpose, as well as their usefulness in the study of disease and the introduction of new concepts and treatments, meant that they were the central sites in which the leading practitioners of medicine congregated.

Though there was already a lying-in ward at the Hôtel-Dieu at least as early as 1664, there were otherwise few facilities reserved for the care of obstetrical patients until the second half of the eighteenth century. Very likely, this situation reflected the general lack of physician interest in the field. But at about this time, the new studies of pathology were producing a vast increase in the understanding of pelvic anatomy and the mechanics of childbirth, so that more and more doctors began to view obstetrics sufficiently challenging for their

attention. Moreover, effective obstetrical forceps, first introduced early in the seventeenth century, were coming into
more common use, and their implementation was considered
a surgical technique. The need for complex knowledge as well
as the growing use of instrumentation demanded a formalized method of training obstetricians, or man-midwives, as
some of them were at first called. A major element in their
acceptance as practitioners of a distinct specialty was the fact
that members of the nobility and upper classes were increasingly turning to them to deliver their children.

Over a period of several decades, therefore, physicians—
all of whom were men—took over control of obstetrics, and
this led to the creation of lying-in hospitals and specialized
divisions in institutions already in existence. The advent of
such facilities raised the status of obstetricians, which had
the effect of making the emerging specialty more attractive
to doctors, thereby increasing the demand for and the creation of more units. This in turn added to the power of
accoucheurs in the profession, although it must be admitted
that it was a long time before obstetrics as an academic field
attained the level long held by internal medicine, and more
recently, surgery.

The founding of lying-in units proved to be a double-
edged sword. The epidemic at the Hôtel-Dieu provided the
first evidence that they would be subject to outbreaks of
childbed fever. In the month of February 1746, twenty puerperal women were taken ill, of whom not a single one sur-
vived. The disease recurred every winter until 1781. (Actually,
there had been an epidemic at the Hôtel-Dieu as early as
1664, but it is poorly documented. There occurred an
unspecified number of deaths, attributed at the time to

impure air arising from the infections of wounded men lying in a ward one floor below.) In a 1790 paper, Joseph Clarke, master of the Dublin Lying-in Hospital (also known as the Rotunda, because there was a circular building on its grounds), reported that "still it prevails more or less in the cold seasons." In 1767, the ten-year-old Rotunda itself had suffered the first of a series of epidemics, which went on until 1788. In fact, between 1764 and 1861, a total of twenty-three separate outbreaks took place at the institution, and the mortality statistics for each of them were grim. During a period in 1770, nineteen of sixty-three women delivered at the Westminster Hospital became infected, of whom fourteen died. Beginning at about this time, one after another institution reported epidemics of the disease, all over Europe. A particularly telling sentence appears in an article about a flare-up of puerperal fever in 1773 at Edinburgh's Royal Infirmary: "Almost every woman, as soon as she was delivered, or perhaps about twenty-four hours after, was seized with it, and all of them died though every method was used to cure the disease. This disease did not exist in the town."

This was an observation that many would make as they witnessed epidemics in all the large cities of Europe. A parturient mother entering a hospital was exposing herself to a mortality rate many times higher than the one she faced at home. Granted, there were sporadic outbreaks in the practice of individual midwives and obstetricians, but the total of these was far less than that of the hospitals. Not only was the incidence lower, but so was the associated mortality: when the disease occurred following a home delivery, 35 percent of its victims died; in the hospital, the figure was usually between 80 and 90 percent.

A particularly striking example of this can be found when statistics of maternal mortality from a facility that performed only outpatient deliveries are compared with those of a lying-in hospital. Thus, between 1831 and 1843, only 10 mothers per 10,000 died of puerperal fever when delivered at home by London's Royal Maternity Charity, while 600 per 10,000 died on the wards of the city's General Lying-in Hospital.

The figures on the Continent were no better. In the decades before and after the turn of the nineteenth century, mortality rates for home-delivered mothers there and in Britain averaged 40–50 per 10,000, while large institutions were reporting figures that were perfectly appalling: at the London General Lying-in Hospital between 1833 and 1842, 587 per 10,000; at the Paris Maternité between 1830 and 1834, 547 (with the highest annual rate during this period being 880, a figure that would later be shown to be at least seventeen times higher than that in the homes of the district in which the hospital was located); at the Dresden Maternity Hospital between 1825 and 1834, 304. These were hardly atypical figures. Nor did Australia and America fare much better. The epidemics experienced in hospitals on those continents may have involved fewer cases and a somewhat lesser mortality, but the New World obstetricians were just as confounded by them as were their counterparts in Europe.

Though seemingly obvious in retrospect, the reasons for the marked differences in incidence and mortality of puerperal fever between inpatient and outpatient populations were not at all clear to the physicians of the time. Hospitals and lying-in units were crowded not only with patients themselves but with accoucheurs and students as well. Internal examinations, many of them having no benefit for the woman in labor, were far more frequent than they were in the home, and often done by young people of little experience and considerable clumsiness. Without knowledge of the role of bacteria in disease, no one gave a thought to the possibility of infectious materials being brought in from the outside, or the cross-contamination that was carried from patient to patient on instruments, linens, and soiled dressings. Infectious material was likely to originate in the delivery units themselves, but another potential source was transmission from the surgical wards, where the mortality from postoperative wound infections approached 50 percent in some institutions. By the 1840s, physicians had come to recognize that puerperal fever was more common when cases of erysipelas were near, but they did not yet understand that transmission from one to the other was the cause of the disease in postpartum women. Least of all were the doctors and nurses thought to be implicated.

In time, there had been so many hospital epidemics of childbed fever that more-careful observations were being made by ever more-skilled dissectors. These increasingly well trained men recognized that the uterus was not always the site of greatest involvement with the inflammation. In some corpses it was the ovaries or tubes that were most affected, or perhaps the large blood vessels entering and leaving the pelvis. In others it might be the omentum, an apron of fat that covers the small intestine. On the basis of its most overt characteristic, physicians began calling the disease by the name of the structure most involved, such as endometritis (endometrium = lining of the uterus), oophoritis (oophoron = ovary), or salpingitis (salpinx = tube). Sometimes it was thought to be primarily an omentitis. When the peritoneum

was the location of most of the process, it was called peritonitis. During the early years of the nineteenth century,
some of the physicians engaged in the new specialty of pathological anatomy began to believe that the originating site of
the disease was not the uterus but its veins and lymphatic
vessels, making it a phlebitis and lymphangitis.

But if this was true, how did the process start? Dr. John
Clarke, a prominent London physician, had expressed the
most commonly held beliefs on this question when he wrote,
in his 1793 text on the management of pregnancy and labor,
"[W]riters think the puerperal fever has evident marks of
putrescency, the cause of which has been traced to miscon-
duct in the early part of pregnancy, such as tight stays and
petticoat bindings, which, together with the weight of the
uterus, detain the faeces in the intestines, the thin putrid parts
of which are taken up into the blood. This is followed by loss
of appetite, in consequence of which bile is collected and
becomes putrid, and is absorbed." As ever, since the days of
Hippocrates, the cause of the disease was thought to be found
in some stagnant or putrefied material whose source was
within the body of the patient herself.

It should not fail to be noted that Clarke referred to the
disease as puerperal, which was a relatively new word at the
time. It was introduced in 1716 by Edward Strother, in a book
called *A Critical Essay on Fevers*, having been derived from
puer, the Latin for "child," and *parere,* meaning "to bring
forth." (*Parere* is, of course, also the origin of "parent.")

Until approximately this time—the early eighteenth century—cases of puerperal fever had been sporadic. Almost all
deliveries were being done in the home, primarily by midwives. But many family physicians were beginning to func-

tion as accoucheurs as well, and were increasingly being called upon when a midwife ran into a complication of pregnancy or childbirth. It was noted by several of these men that from time to time, either he, a local midwife, or a fellow practitioner might experience a string of several deaths in succession. Periodically, there might even be an outbreak of the disease lasting for months or longer, restricted to the patients of one attendant. It did not escape some of the more thoughtful of such doctors that the single characteristic shared by the affected mothers was that they all were delivered by the same person or persons. Out of this observation began to appear the first inklings that perhaps there was a causative factor coming not from the women themselves but from some external source common to each of the victims, namely the accoucheur.

But this was a serious accusation, one that could blight or even ruin the career of the doctors or midwives so charged, not to mention the agonies of conscience they might undergo. Anyone suggesting such a thing would need to have an experience of the disease so wide that the conclusion was tenable, and also the stout heart to declare it in medical meetings and literature. And in that era of multiple confused theories not only about the origins of puerperal fever but in the entire greater arena of medical practice itself, it would take a particularly well educated and intellectually sophisticated physician to interpret the observations correctly and bring all the evidence together into a coherent whole. After all of this, he would have to face his colleagues armed with the courage of his convictions.

Such a man was the Scottish physician Alexander Gordon. Gordon, born in 1752 in a town near Aberdeen, had learned his profession at the two most eminent medical schools of the period, those of Edinburgh and Leiden, where he went for further study following a period of clinical instruction at the Aberdeen Infirmary. After spending five years as a surgeon's mate in the Royal Navy, he moved to London in order to train under several of the most prominent obstetricians of the day. Returning to Aberdeen, he founded a free dispensary where he gave instructional courses to the midwives of the city and began a private practice in all aspects of medicine. He was not only skilled in the management of the entire spectrum of disease, but he was the only formally trained obstetrician in the city.

In December 1789, barely four years after Gordon had established himself in Aberdeen, an outbreak of puerperal fever occurred there, which lasted until March 1792. Virtually all of its victims came under his direct or indirect scrutiny, and he seized the opportunity to keep careful records of his observations. He was struck by similarities he detected between the course of the disease and that of erysipelas, the name given since Hippocratic times to a rapidly spreading inflammation or infection, which, by seeming coincidence, raged in Aberdeen precisely at the same time as did puerperal fever. The infections were so prevalent, Gordon noted, that almost every injured patient, male or female, admitted to the Aberdeen Infirmary "was soon after his admission seized with erysipelas in the vicinity of the wound."

To Gordon's perceptive eye, this observation gave the lie to any puerperal fever theory based on lochia, milk, or the variety of fanciful notions then being accepted by the medical establishment. It contradicted also any thought that there might be validity in the concept that others called "the epidemic constitution," believed by many to be the factor that made some women more susceptible than others: the seventy-seven patients he personally treated (of whom twenty-eight died) were a cross section of the population of Aberdeen, transcending boundaries of income, temperament, marital status, age, stature, and robustness. Moreover, he found no evidence that some noxious state of the atmosphere was enveloping individual women or the city, a condition called miasma, invoked since ancient times to explain otherwise unexplainable epidemics.

Gordon became convinced that puerperal fever was contagious. But his imagination did not reach to the point of postulating that some specific agent was being transmitted. He restricted himself to the speculation that "every person, who had been with a patient in the Puerperal Fever became charged with an atmosphere of infection, which was communicated to every pregnant woman, who happened to come within its sphere." But he had no doubt that the method of spread was contagion. In 1795, the forty-three-year-old physician published a seven-chapter book, *A Treatise on the Epidemic Puerperal Fever of Aberdeen,* in which his formulations were lucidly explained, with the support of tables, autopsy reports, and other verifying data. The crux of his argument was as follows:

That the cause of this disease was a specific contagion, or infection, I have unquestionable proof. . . . [It] was not owing to a noxious constitution of the atmosphere . . . for, if it had been owing to that cause, it would have seized women in a more promiscuous and indiscriminate manner. But this disease seized such women only, as were vis-

ited, or delivered, by a practitioner, or taken care of by a
nurse, who had previously attended patients affected with
the disease. . . . [T]he infection was as readily communicated as that of the small-pox, or measles, and operated
more speedily than any other infection, with which I am
acquainted. . . . It is a disagreeable declaration for me to
mention, that I myself was the means of carrying the infection to a great number of women.

He went so far as to say, "I plainly perceive the channel by
which it is propagated; and I arrived at that certainty in the
matter, that I could venture to foretell what women would be
affected with the disease, upon hearing by what midwife they
were to be delivered, or by what nurse they were to be
attended during their lying-in: and, almost in every instance,
my prediction was verified."

Gordon's recommendations for preventing further epidemics and individual cases were similar to those already being
used when there was danger of serious contagious diseases, like
smallpox. Chief among them were fumigation of rooms and
beds, burning of nightclothes and bed linens, and scrupulous
washing by physicians and nurses who had been exposed, along
with fumigation of all apparel worn by them. As for treatment,
his belief in heroic bleeding of twenty to twenty-four ounces
(to be repeated if no improvement occurred) was greeted with
skepticism, though he claimed much success with it.

It was for good reason that Gordon referred to his discovery of the contagiousness of puerperal fever as "the fatal
secret." Not only did he make the mistake of publishing the
names of seventeen involved midwives, but his admission of
his own culpability virtually assured him the ire of female

accoucheurs and the population of Aberdeen at once, not to mention the few of his colleagues who delivered babies. The commotion proved too much for Alexander Gordon. In a short time, he departed Aberdeen and reenlisted in the navy. He died of tuberculosis at forty-seven, having severed all of his connections to obstetrics.

Gordon's relationship to obstetrical theory ended almost as unobtrusively as did his career. His propositions, which had never been widely known, were forgotten by all but a few serious students of the problem, not to be generally recollected until the late nineteenth century, when the principles of germ theory so well documented by such as Joseph Lister, Louis Pasteur, and Robert Koch were being gradually accepted as the explanation for contagious disease. It was Gordon's fate to be neither a herald nor heralded.

Gordon's obscurity to his own and succeeding generations is all the more puzzling in view of the increasing experience of outbreaks of puerperal fever that physicians encountered in their practices. With the passage of time in the early nineteenth century, more and more men were devoting the majority of their time to obstetrics, becoming well known as proficient accoucheurs and building large followings of women. Inevitably in the practice of some of them, there would now and then occur an unexplainable outbreak of childbed fever, sometimes affecting one woman after another while patients of other physicians in the same area went unscathed.

Even the few physicians who had taken Gordon seriously and continued to remember his work had difficulty accepting the possibility that they or their colleagues were the proximate cause of such devastation. Some of them nevertheless took great pains to disinfect themselves after a puerperal fever death, while others took equally great pains to deny that medical attendants might be transmitting the disease to their patients. Charles Delucena Meigs, for example, professor of midwifery and the diseases of women and children at Philadelphia's Jefferson Medical College and perhaps America's most highly respected obstetrician, was a vocal opponent of such a formulation, citing example after example of the presumed lack of association. Of the cases of puerperal fever he saw in his practice, he famously said, "I prefer to attribute them to accident, or Providence, of which I can

Charles Delucena Meigs. (Courtesy of the National Library of Medicine)

form a conception [*sic*], rather than to a contagion of which I cannot form any clear idea, at least as to this particular malady." In his popular 1851 textbook, *Woman; Her Diseases and Remedies,* Meigs told a story that would be amusing, were its outcome not tragic, of a colleague whose patient died of puerperal fever despite every conceivable precaution on the part of the attending obstetrician. The colleague, one Dr. Rutter of Philadelphia, had the unfortunate experience of treating forty-five cases of puerperal fever in 1843 and twenty-five in 1844. Believing himself to be the source, Rutter did everything he could to prevent further contagion, but not always with success. Dr. Meigs used the following description to argue his case against the hypothesis of transmission by doctors:

> My friend Dr. Rutter informs me, that to one of the cases, he was summoned on the night of his return to the city, after an absence from it of ten days, at a distance of thirty-five miles. Previously to visiting the patient, he entered a warm bath, had his head shaved, put on a new wig, new hat, new boots, and pocket-handkerchief, and every article of his dress was bought new for the occasion; leaving at home even his watch and pencil, and taking care, after the bath, not to touch a single article of the clothing he had previously worn. The patient whom he had attended was immediately seized with the symptoms of puerperal fever. I was called to see her along with him, and attended her up to the period of her death, which took place on the eleventh day after the birth of her child.

The unhappy scenario occured in 1844, but Meigs does not provide the month. One cannot help wondering whether it was near the end of the year, for only six of Dr. Rutter's patients developed the disease in 1845 and seven in 1846, after such a long period of grievous outcomes. Perhaps the bathing and the shaving and the new clothes did some good after all. There have always been, and will always be, cases of childbed fever—though not many—that arise from some cause of contamination other than the medical attendants themselves.

A few years before Rutter had his unfortunate experiences, a New England physician in his early thirties had attended a meeting of the Boston Society for Medical Improvement during which he partook in a discussion "of a certain supposed cause of disease, about which something was known, a good deal was suspected, and not a little feared."

> The discussion was suggested by a case, reported at the preceding meeting, of a physician who made an examination of the body of a patient who had died of puerperal fever, and who himself died in less than a week, apparently in consequence of a wound received at the examination, having attended several women in confinement in the mean time, all of whom, as it was alleged, were attacked with puerperal fever.

Gifted with an insatiable curiosity that would serve him well in his subsequent career as an anatomist, and a deep sympathy that would serve him equally well in his equally successful later career as a poet and essayist, the young physician, Oliver Wendell Holmes, undertook a study of the problem of the transmission of puerperal fever. He reviewed numerous records, read countless medical articles and books—including Gordon's—and consulted many colleagues. He did not start his investigation in a vacuum of informed opinion, though. As he would point out in his report, there was already plenty of evidence, and many doctors were beginning to agree, that puerperal fever is a contagious disease. His intention was to marshal the observations in such a way as to guarantee that "no man has the right to doubt it any longer." Specifically, he aimed to persuade or debunk the contrary opinions of some of the era's leading obstetricians, who, in spite of what Holmes considered incontrovertible proof, persisted in error. He later wrote, "[A] few writers of authority can be found to profess a disbelief in contagion—and they are very few compared with those who think differently." Chief among the "writers of authority" was Charles Delucena Meigs.

At the end of his exhaustive quest, Holmes came to conclusions that were amply supported by all the information he had been able to gather. They can be summarized in a single sentence, though that sentence does not sufficiently convey the persuasiveness of Holmes's evidence, the forceful clarity of his argument, or the elegance of his literary style: *The disease known as Puerperal Fever is so far contagious as to be frequently carried from patient to patient by physicians and nurses.* The italics are Holmes's.

Like Gordon before him, Holmes did not profess to know the mechanism whereby medical attendants spread the disease, but his choice of words could, perhaps, be interpreted to mean that he was raising the possibility:

I shall not enter into any dispute about the particular *mode* of infection, whether it be by the atmosphere the physician carries about him into the sick chamber, or by the direct application of the virus to the absorbing surfaces with which his hand comes into contact. Many facts and opinions are in favor of each of these modes of transmission. But it is obvious that in the majority of cases it must be impossible to decide by which of these channels the disease is conveyed, from the nature of the intercourse between the physician and the patient.

Of course, the word "virus" did not have the meaning with which we today associate it. Derived from the Latin word for "poison," it was a general term referring (as described by a medical dictionary of the time) to "a principle, unknown in its nature and inappreciable by the senses, which is the agent for the transmission of infectious diseases." Thus, by his use of "virus," Holmes was not, in fact, implying that childbed fever was caused by a particle or organism.

But Holmes did have strong recommendations to make, which, if followed to the letter, promised to decrease markedly the incidence of the disease: physicians should avoid autopsies on cases of puerperal fever or erysipelas when preparing to attend at a delivery; if it is absolutely necessary to be present at such an autopsy, all clothing must be changed and a twenty-four-hour period should elapse before any delivery is undertaken; upon discovering a case of puerperal fever in his practice, a physician is obliged to consider his next patient in danger of becoming infected, and take appropriate precautions; any physician experiencing two cases within a short interval should relinquish his practice for at least a month; and doctors must take every care to be certain the disease is not being transmitted by nurses or other assis-

tants. Though Holmes did not advocate any specific form of
disinfection of the body other than washing, he pointed out
in later editions of his treatise that one "Semmelweiss" had
experienced an "[a]lleged sudden and great decrease of mortality from puerperal fever" by using a nailbrush and disin-
fecting his hands with chloride of lime.

But Holmes at the time of his publication was a young man
at the very beginning of his career, and not even an obstetrician at that. The forces of which Charles Meigs was the proto-
type attacked him with all the authority and verbiage at their
command, made all the more determined because Meigs
himself was singled out by Holmes's lance of logic. The
Philadelphia professor dismissed his young adversary as an
amateur and called his study "the meanderings of a sopho-
more." Even less vocal opponents were skeptical, and believed
the Holmesian doctrine to be unproved, questioning his assertion that most "writers of authority" agreed with him. In addition, Meigs and the others may have been offended at the
boldness of a young man who closed his presentation with an
indictment of those who might be expected to ignore his thesis. There did indeed seem more than a whiff of arrogance in
an as yet unproven author who had the temerity to claim not
only that he had provided the ultimate proof to the solution of
a two-thousand-year-old mystery but that those who did not
adhere to his principles were felons. As the thirty-three-year-
old doctor charged in the uncharacteristically intemperate final
paragraph of the paper he read at the February 13, 1843, meeting of the Boston Society for Medical Improvement,

> Whatever indulgence may be granted to those who have
> heretofore been the ignorant causes of so much misery,

the time has come when the existence of a *private pesti-lence* [italics his] in the sphere of a single physician should be looked upon, not as a misfortune, but a crime; and in the knowledge of such occurrences the duties of the prac-titioner to his profession should give way to his paramount obligations to society.

Holmes's zeal did not endear him to his contemporaries, at least not to those outside of Boston who were not familiar with the sterling qualities of character and intellect that were already being recognized in his native city. The response of Meigs to this statement (and especially to the word "crime") cannot be known, but he would certainly have found it even more offensive than the rest of Holmes's treatise, which was soon to be published as an essay destined to become famous under the title "The Contagiousness of Puerperal Fever." Historians are fond of heaping scorn on the dean of American midwifery for his stubborn refusal to accept the notion of transmission by medical attendants, but he does deserve more credit—and even a bit of redemption—than most have been willing to grant him. In the chapter on puer-peral fever in *Woman; Her Diseases and Remedies,* published when he was almost sixty years old, Meigs delivered himself of a long argument against theories of contagion and then had this to tell his readers, much as he had for years been say-ing it in lectures to his many students:

From all the foregoing, you will perceive, my young friends, that I assert my disbelief in the contagiousness of puerperal or childbed fever; and you will have noticed at the same time that, notwithstanding my plenary assertion

of this disbelief, I am not able to fly in the face of the asser-
tions, and opinions, and sentiments of many of my medical brethren, worthy of my highest respect; so that, in fact,
I do not feel at liberty to disobey their injunctions, at taking all proper precautions against propagating by my person a malady so fatal in its nature. And I therefore most
explicitly declare, and I beg you to bear in mind that I now
make this declaration, that I think it will always be your
duty, whether you may believe in the contagiousness or
not, of a malady, to avoid, as far as it may be in your power,
all occasion to transmit, if it can be transmitted, an epi-
demical or endemical disorder.

In other words, though he did not believe in contagion,
Meigs taught that doctors should always conduct themselves
as if the theory were correct. He was like an atheist who
prays—"just in case."

III

In spite of its later obscurity, Alexander Gordon's work was at first known by many English-speaking obstetricians, but never taken very seriously on the Continent. In Britain, others had begun to write of much the same observations as his, and the theory of contagion was accepted by enough of them that they became known to authors on the subject of puerperal fever as "the English contagionists." But theirs remained a controversial notion, debated at medical associations and in the pages of journals. Not only Charles Meigs but other influential obstetricians as well set no store by it. He and those who shared his views became known as anticontagionists. Some who took a middle ground were called contingent contagionists, signifying their belief that it might or might not be tenable in individual instances, depending on the circumstances in which the disease occurred. The contagionists' major contention, that some sort of aura around the accoucheur was being carried from patient to patient, was unacceptable to the majority of physicians, and yet the medical literature of the time is filled with descriptions of physicians—almost always those who did their deliveries in patients' homes—burning their clothes, fumigating their instruments, shaving their heads, and performing similar contortions when confronted with an epidemic in their private practice.

Still others subscribed to a thesis promulgated in 1773 by Dr. Charles White of the Manchester Infirmary, according to which the instigating factor was foul air. While accepting the commonly held belief that tight stays, bindings, and petticoats caused retention of feces, which were then absorbed into the circulation, he added another condition, which he considered crucial. In describing home deliveries, he wrote of what he called "a putrid atmosphere" in which many women give birth:

> When the woman is in labour, she is often attended by a number of her friends in a small room, with a large fire. . . . [B]y the heat of the chamber, and the breath of so many people, the whole air is rendered foul, and unfit for respiration; this is the case in all confined places, hospitals, jails and small houses, inhabited by many families, where putrid fevers are apt to be generated, and proportionally the most so where there is the greatest want of free air. Putrid fevers thus generated are infectious.

To combat the closeness of such an atmosphere, White proselytized for free ventilation by means of such measures as the opening of windows and the maintenance of wide-open chamber doors. In no way did he associate disease with the presence of doctors, or even midwives. When Joseph

Clarke was master of the Rotunda, he went even farther to combat the string of epidemics. By his order, holes were bored in sashes and doors, and the ceiling of every ward had a 24-by-6-inch opening; the windows were equipped with louvered panes. The "bad air," so commonly indicted as a cause of sickness, was literally bad air, but in the conception of those who shared the view of White and Clarke it was the result of poor ventilation. Always, there is the notion of something noxious in the atmosphere, as though it were a localized miasma.

When doctors of the period prior to the promulgation of the germ theory (in the 1870s and 1880s) used such words as "infectious" and "contagion," they thought they understood what those terms meant, at least in the general sense. But with the wisdom of a later time we know that they were mistaken. As late as 1874, a physician consulting Robley Dunglison's *Dictionary of Medical Science,* a popular American reference work, for a definition of "infection" was referred to "contagion," where he would find the following:

> The transmission of a disease from one person to another by direct or indirect contact. The term has, also, been applied by some to the action of miasmata arising from dead animal or vegetable matter, bogs, fens, &c, but in this sense it is now abandoned. Contagious diseases are produced either by a virus, *contagium,* capable of causing them by inoculation, as in small-pox, cow-pox, hydrophobia, syphilis, &c, or by miasmata, proceeding from a sick individual, as in plague, typhus gravior, and in measles and scarlatina [?]. Scrofula, phthisis pulmonalis, and cancer have, by some, been esteemed contagious, but apparently

without foundation. Physicians are, indeed, by no means unanimous in deciding what diseases are contagious, and what not. The contagion of plague and typhus, especially of the latter, is denied by many. It seems probable that a disease may be contagious under certain circumstances and not under others.

(The bracketed question mark is in the original text. Also, it should be remembered that the notion of "virus" was a general concept at that time. The word was used to refer to any unspecified principle or agent that is capable of causing transmissible disease. The possibility of a fragment of organic matter of some sort was considered by an occasional author, but seems rarely to have been given much credence except in the highly theoretical sense. For example, a Jesuit priest named Athanasius Kircher had in 1658 suggested that "small, living animals invisible to the naked eye" spread contagious disease, but he was ignored by physicians of later centuries. Long before Kircher, Girolamo Fracastoro of Verona had written a book called *De Contagione* in 1546, suggesting the same thing. Such seers were unremembered, or their writings were considered historical oddities by the up-to-date physicians of the eighteenth and nineteenth centuries. At best, they were thought to be interesting scribblings, of no more consequence to contemporary medicine than the tomes of medieval scholars.)

This is a long entry, but one worth pondering. In a single paragraph Dunglison presents an epitome of the confusion that existed even half a century after the events being described in these pages. There is a deliberate vagueness to it, which would not be dispersed until the germ theory had become a prominent factor in medical thought during the late nineteenth century. No wonder that the concept of contagion was so often misunderstood, and the word itself used in connotations that added to the uncertainties. And no wonder also that in reviewing the events of these times, a modern reader is so easily thrown into a quandary in trying to interpret small shadings that differentiate one usage from another. Specificity of meaning would come to this and other areas of medical theory only when science itself became more specific and a consistent vocabulary could be developed.

But one consistency in beliefs about the propagation of puerperal fever was that it was attributable in some way to a kind of noxious atmosphere to which a patient became exposed. Accoucheurs who delivered at home gradually came to accept that they themselves might be the source of the infection, by carrying the effluvium on their clothes or person, or perhaps in the air surrounding them. They were the ones most likely to take the precautions noted earlier, of bathing, fumigating, and closing their practices for periods of time. But such realizations seemed rarely to strike hospital obstetricians, very likely because the epidemics they witnessed involved patients of every doctor in the institution, rather than those of one individual. They concentrated their efforts at prevention on the lying-in units themselves, including walls, bedding, and instruments.

Or at least, obstetricians believed, there was an influence at work in the surroundings in which the epidemic took place. Sometimes the indicted influence was something as simple as defective drainage or the existence of nearby foul sewers. Some seized on the seeming relationship of outbreaks to seasons of the year or weather to invoke telluric or cosmo-telluric or even solar or magnetic influences. But though there were disagreements about its origins, virtually everyone who thought carefully about the disease seemed to subscribe to the notion that the presence of so much as a single case created some sort of effluvium that had to be dispersed if further sickness were not to occur. The notion of miasma lurks within each theory.

Various means were utilized to rid hospital wards of the miasmatic influence. Most of them amounted to methods that we nowadays recognize as disinfection. Wards were closed, concentrated chlorine gas was pumped into them for a day or two, the walls were covered with a cream containing calcium hypochlorite (chloride of lime), all surfaces were freshly painted, beds were replaced, linens and blankets were put into a high-temperature stove, apparel and bedclothes were burned or thoroughly fumigated, fires were lit in the wards and their smoke allowed to permeate every cranny of it, and so on.

Sometimes these measures succeeded. When the ward was reopened, there was no puerperal fever for a while. But in time, the cases would always recur, and another epidemic was sure to follow. In the long run, every attempt to end the ravages of the disease failed, because no one recognized its true source. Puerperal fever was being caused not by an atmosphere, a miasma, or any other such vague aura—or by milk metastasis, lochial suppression, cosmo-telluric influences, personal predisposition, or any of the other chaos of supposed factors—but by direct inoculation from the hands of the very physicians battling to prevent its occurrence. In order for that to be demonstrated, an entire sequence of events was required that would change the direction of medical thought and alter the ways in which disease was investigated and understood. The movement had its beginnings in the final third of the eighteenth century, and was well under way during the worst period of the hospital epidemics. It found form in a new approach to the study of sick organs and tissues, called pathological anatomy.

Pathological anatomy is a field of inquiry that lies at the basis of all scientific medicine. Most simply defined, it is the study of the structural changes that occur in organs and tissues when they become diseased. Along with the overlapping field of pathological physiology (which arose later, to study abnormalities of *function*), it is the primary route toward a comprehensive understanding of the body's mechanics, whether sick or healthy. The great mass of our knowledge of human biology has been obtained by investigating abnormalities of structure or function. Tracking of the telltale clues discovered in such studies leads to an understanding of the physical and biochemical processes that underlie life. Whether one wishes to call it the cornerstone of medicine or its foundation, the investigation of pathological structure and function is the basis of virtually all we have been able to learn about our bodies.

The field of pathological anatomy may be said to have been founded in 1761 with the publication of a single text, written by a researcher who had spent more than five decades in its preparation. Giovanni Battista Morgagni, professor of anatomy at the University of Bologna, was seventy-nine years old when he completed *De Sedibus et Causis Morborum per Anatomen Indagatis* (On the Seats and Causes of Disease, Investigated by Anatomy), a contribution whose title summarizes its landmark message: If you want to understand disease, you must identify its seat, the place in which it originates. Symptoms are, Morgagni famously proclaimed, "the cries of the suffering organs," and they must be traced back to the specific locations that produce them. *Ubi est morbus?* was the question to be answered in each case, "Where is the disease?" This was the question Morgagni posed to the ages. It would become the unspoken motto for every succeeding generation of medical scientists and clinicians.

To us as twenty-first-century beneficiaries of the subsequent two hundred years of medical progress, the need to answer that question would seem to be self-evident, even banal. How, after all, can the causes of a disease be understood if the precise location from which it exerts its baleful influences is not known? How, moreover, can a diagnosis be made if the physician cannot trace symptoms back to their organ or tissue of origin? How, finally, can treatment be devised if it cannot be directed to the specific organ or group of cells whose disordered state is to be corrected? Our very nomenclature of disease is to a great extent founded on acknowledgment of the critical importance of identifying sites of abnormality. When we refer to a sickness as appendicitis, duodenal ulcer, or myeloma, we are indicting the organ or tissue in which we know the morbid condition makes its home. These diseases were neither named nor even recognized prior to the development of the discipline of pathological anatomy.

Obvious as it now seems, locating the site at which symptoms originate was of little consequence to physicians prior to 1761. Since long before the days of Hippocrates, sickness had been thought to be caused by generalized changes within a patient's entire constitution, reflecting some outside influence to which he had succumbed. On the most primitive level, it was due to punishment by the gods or God for some miscreance (a notion from which individuals among us, incidentally, are not yet entirely free). When Greek physicians began codifying medical concepts around four centuries B.C.E., they formalized the old Egyptian notion of the balance of several internal fluids, or humors, necessary to maintain the equilibrium of health. The Hippocratic authors wrote of blood, phlegm, yellow bile, and black bile, influenced by such factors as surroundings, life events, climate, the seasons, and state of mind. Diseases were believed to be generalized indispositions, to be treated with generalized methods meant to restore the generalized balance. The specific location in which the problem manifested itself was of little importance, and there was no advantage in identifying it. Therapies were focused on a patient's entire constitution. The most frequently used correctives were those directed at eliminating a surfeit of one or another humor thought to be causing the problem. Accordingly, bleeding, purging, puking, and sweating were among the mainstays of the doctor's armamentarium, to drain or force out the excess. Massages, bodily exercises, and changes of diet or surroundings were often prescribed to complement these methods, as were various forms of baths, from mud to sun.

As the centuries passed, layers of theory were added to these basic principles, but the underlying philosophy remained the same, namely that disease was constitutional and that its treatment must be directed to restoring the balance of the entire body. Complex etiologies in time came to be evoked in order to explain maladies that struck not only individuals but also affected groups of people and even entire populations. Medical theorists wrote of such abstruse epidemic factors as telluric influences, those arising from the earth itself. Medieval, Renaissance, and later literature is replete with references to them as well as to those that are cosmologic, magnetic, or a combination of all or several of them. And there persisted the ancient notion of miasma, arising from the earliest mists and myths of medical philosophy. A miasma can best be described as an emanation or an atmosphere, whether from the earth itself or from some particular area, that hovers in the surroundings and causes sickness in those exposed to it, by the pervasiveness of its malign presence. In keeping with such theory, miasma was one of those generalized influences that cause a generalized disease. It was to be treated with generalized methods. These notions were reinforced by certain easily observed realities, such as the prevalence of malaria (the word literally means "bad air") in places of stagnant water and marshes, with the resultant pollution that was thought to hang on everything in the vicinity.

Diagnosis and treatment were ruled by such formulations. From time to time, the body of a patient would be opened soon after death and some more specific cause of disease looked for, but no systematic studies were undertaken. Such individual investigations became more common during the scientific revolution of the seventeenth century, a time when specificity of cause and effect became a hallmark of the emerging process of experimentation. Finally, a Swiss physician named Theophilus Bonetus was able to assemble reports of some three thousand such autopsies by 470 authors, which he published in the form of a book in 1679. But Bonetus was so disorganized in his approach that doctors attempting to understand his message found it to be incoherent. Morgagni originally intended merely to clarify and rewrite his predecessor's work, but he soon realized that an entirely new study was required. To produce it became his lifework. Because he was not only a skilled anatomist but a talented bedside physician as well, he was able to make observations in his own autopsy series and those of others that might have eluded less well prepared thinkers. Cross-referenced and multiply indexed, Morgagni's opus enabled any reader to link a patient's complaints and appearance to the organic changes of which they were the result. He was the first to make a real correlation between the bedside doctor's clinical observations and the gaze of the postmortem dissector. In that single five-book tome of some seven hundred case histories and subsequent autopsies, he created the field of pathological anatomy.

De Sedibus epitomized Enlightenment ways of studying the human condition. At a time when thinkers were questioning every given that had been bequeathed to them, Morgagni brought forth an immense and encyclopedic work that shook the foundations of the entire edifice of speculations that had until then characterized physicians' approaches to disease. A new spirit of medical inquiry arose, in which the catalytic factor became the Morgagni-inspired zeal to study the finest details of case histories with an eye to discovering at subsequent autopsies the underlying organic pathology causing the evolution of the process that had been encountered at the bedside.

The fresh intellectual winds released by the events of 1789 swept the cobwebs from the eyes and minds of French physicians and gave them the leadership of the new medicine. The hospitals of Paris became the centers of the new philosophy of discovering "the seats and causes of disease, revealed by anatomy." Students and senior physicians alike gathered from all parts of Europe and America to trail after the emerging cadre of Parisian master clinicians on their rounds and then to the autopsy rooms. A patient seen on the wards yesterday was autopsied today, and the reasons why the suffering organs cried were made clear. The autopsy became the key to understanding not only disease but its evolution in the previously healthy men and women it had killed. The visiting doctors returned to their home countries imbued with the new philosophy of medicine, epitomized by the autopsy and the clinicopathological correlation.

Methods were soon devised that might enable a physician to track symptoms to their organs of origin while the sick man or woman still lived. Doctors began to look more carefully at the appearance of a patient, finding evidence of disease in such visible factors as gait, skin color, and surface irregularities; they attempted to feel certain organs through their soft coverings, as was possible with some of the structures in the neck and the abdominal cavity. They called these new diagnostic procedures "inspection" and "palpation." It was discovered that tapping on one outstretched finger of a hand pressed up against the chest gave audible clues to the degree of normality of the underlying lung, and to the presence of pathological fluid. The doctors who used this technique called it "percussion." In a time rampant with tuberculosis and other pulmonary diseases, many clinicians identified abnormalities by placing an ear directly on the chest wall and listening to the quality of a patient's breathing; or they might detect iregularities of the heart in the same way. This clinical listening became known as "auscultation," which evolved into a fine art after the stethoscope was invented in 1816 by a Frenchman named René Laennec.

Thus were developed the basic principles of the physical examination, with its four components of inspection, palpation, percussion, and auscultation—all of which came into use shortly after the turn of the nineteenth century in Paris. Specificity of diagnosis became the holy grail of medical practice, in which the ultimate proof of one's accuracy lay in the autopsy. It was an exciting time, as correlations between physical findings and pathological anatomy were constantly being discovered. The basis of everything was the well-conducted dissection of the patient's cadaver.

In no other part of the world was the cause of autopsy so forcefully taken up as in the German-speaking countries. By the middle of the nineteenth century, as a result, the hospitals of Germany, Austria, and Switzerland had snatched medicine's leadership from the French. Their ascendancy was based largely on their enthusiastic embrace of the notion that they could find clues to human biology in the laboratory, by studying the specific organs and tissues to which disease was traceable. They did this by means of autopsies even more detailed than those of the French, and by the investigation of disordered physiology. In order to study the new pathological physiology, the field of biochemistry would later be developed in these countries, thus setting the stage for the astonishing discoveries that eventually led to the all-embracing approach to health and disease that in the middle of the twentieth century became known as biomedicine.

Of all the German-speaking academic institutions, none was more scientifically advanced by the mid-nineteenth century than the medical school of the University of Vienna.

Great as its reputation had already become, a change made in its faculty in 1844 was destined to burnish its reputation even further and have far-reaching consequences not only for the institution itself but for the course of medical thought throughout the world: on the basis of his already valuable contributions to the understanding of the processes of disease, the forty-year-old Karl von Rokitansky was elevated from the position of associate professor to become the director of pathological anatomy.

The study of pathological physiology still being barely in its earliest period of gestation at the time of Rokitansky's promotion, the focus remained entirely on pathological anatomy. At the Allgemeine Krankenhaus, as at all public German and Austro-Hungarian hospitals, every patient who died was taken to the autopsy room for postmortem examination. Countless observations were being made, recorded, and studied by the meticulous Germanic dissectors, and large numbers of books and articles emerged from that crucible of medical investigation. In 1844, to be an academic physician at one of the Germanic universities was to be convinced that all worthwhile knowledge radiated from the great hospitals of Vienna, Berlin, Zurich, Prague, and similar cities.

There is no better description than Rokitansky's own of the method of these dissections and its intent. Rokitansky was not only Europe's—and the world's—leading exponent of this approach but its founder. Of his procedures, he would later write in his autobiography, "First . . . sorting the facts scientifically on a purely anatomical basis . . . second demonstrating the applicability of the facts and their utilization for diagnosis in live patients." Rokitansky's plan was to seek patterns in the thousands upon thousands of observations he made so that categories of disease might be established that would allow classification and yield clarity of the distinct entities. So successful was he in this endeavor that Rudolf Virchow of the University of Berlin, the next generation's leader in the field, called him "the Linnaeus of pathological anatomy."

Rokitansky's accomplishment was to take the initiative introduced by the French and bring it to maturity. Whereas in the unfocused situation at the Paris hospitals—each autopsy was done by the clinician who had treated the patient—in Vienna Rokitansky himself did all dissections, a practice he had begun long before assuming the chair. With such a centralization in his expert hands, one after another publication came forth, establishing not only the new basis of medical theory but also promoting the role of the laboratory in the study of disease. His multivolume *Handbuch der pathologischen Anatomie,* published in a series during the 1840s, became the bible of the movement he was spearheading, and assured the position of the Allgemeine Krankenhaus as a Mecca to which physicians made pilgrimages from all over Europe, Asia, and America. As Erna Lesky, the eminent historian of medical Vienna, would write more than a century later, "Vienna medicine had become world medicine. Rokitansky stood at the height of his fame" at the time of his promotion in 1844. In that year, pathological anatomy became a compulsory subject at the school.

No longer was disease to be studied only at the bedside—a patient's corpse was now seen as a treasure trove of enlightenment, to be scrutinized without the hurry required by the urgent needs of the sick. Within a few decades, microscopy and chemistry would advance to the point where body fluids and tissue samples were added to the sources of information. In this, too, the German laboratories would be the leaders. Rokitansky personally conducted some thirty thousand autopsies during his long career, and the benefits to medical science were incalculable.

Of all the diseases studied by the dissection of the dead at the Allgemeine Krankenhaus and elsewhere, none was more mystifying than childbed fever. Thousands upon thousands of autopsies had been done on the bodies of its victims; every aspect of the postmortem findings was well known, and yet the cause remained a mystery. Convinced that the solution to the riddle lay in even more autopsies, doctors continued to dissect the corpse of every mother whose body came into their hands. And, thanks to the large number of hospital epidemics, there were plenty of them.

Rokitansky trained many able disciples, who came to the medical school from every part of the far reaches of the Austrian Empire. Some of them would make contributions to medical science almost as great as his own. But only one turned his attention wholly to the problem of puerperal fever. Using the methods taught to him by his great master, that young man, Ignác Semmelweis, did in fact find the correct answer to the problem he set out to solve. But instead of gaining fame and the plaudits due a benefactor of humanity, Semmelweis was destined to see his dreams destroyed and his health ruined. It is to his remarkable story that this narrative must now turn.

IV

Nothing in the background of Ignác Semmelweis gave the slightest indication that he would spend the greater part of his career a rebel against conformity. In fact, a life of conformity seemed to be precisely what he was destined for. There is no hint of defiance or disobedience to the established order in any of the identifiable generations that preceded his birth on July 1, 1818, as the fifth of nine children of the grocer József Semmelweis and the former Terézia Müller. József's father, János, had been a vineyard hand in the Hungarian village of Kismarton; János's father (another János), was probably a farmer; and this János's father, György, a farmer at Szikra. The study of parish registers has made it possible to trace the family back to the small village of Marczfalva in 1570, and no whisper of dissension has appeared in any archive of its history. The family's tenure in Budapest began in 1806, when József, still a bachelor, moved from Kismarton to the thriving capital of Hungary, then known as the "free, royal city of Buda."

József became no ordinary grocer. By marrying Terézia, the daughter of a wealthy Bavarian-born coach manufacturer, he added affluence to the personal qualities that assured the growth of his reputation and business, which, some years later, allowed him to achieve the position of *Obervorsteher*, or chairman, of the Grocers Association. In time, he would own four houses, including the one on Apród Street in the Tabán district, where his shop and the dwelling place of the family were located.

The histories of the four eldest Semmelweis children have been traced, and the results of the search are not surprising: they all lived their lives in ways of which their parents would have approved. Not unexpectedly, József, the first son, went into the family business, eventually owning a large grocery store opposite the medical school of the University of Pest; next in line was Károly, who made his father and mother proud by taking holy orders as a Roman Catholic priest; following him was Fülöp, a tradesman like his father and brother; then came Julianna, the first daughter, whose pharmacist husband inherited his father's chemist's shop. Only Ignác shook off the traces that seem to have been so carefully put into place by Terézia and József.

The Semmelweis family was in every way part of the bourgeoisie, with particular emphasis placed on the values of that class by a father newly entered into it. Probably descended from a line whose origin was Frankish, József, like most of his friends and colleagues in the trading population of Buda and Pest, spoke a Germanic dialect called Buda-Swabian, both at home and in his business. Because his wife had a Bavarian father, her household conversations were in German. The result was that young Ignác—or Naci, as he was called by his relatives—did not learn to use the Hungarian language properly until secondary school. His speech would be inflected with a distinctly regional Germanic accent for the rest of his life, reflecting the Buda-Swabian atmosphere that surrounded him in his childhood and adolescence. During his later years in Vienna, the sound of his voice betrayed his origins.

Ignác was enrolled in the Catholic Gymnasium at the age of eleven, and the records indicate that he was an industrious and able student throughout, though the academic standards of the country's secondary schools were not high. German and Latin were well taught, but Hungarian much less so. The result was that few students had learned to read, write, and speak the language correctly by the time they graduated. Though Ignác achieved excellent examination grades, there is no reason to believe that he was in this regard any different from the usual product of these schools. Even having been awarded second-level honors at graduation, he was hardly skilled in the native language of his own country. Moreover, as he himself would point out decades later, he had developed "an innate aversion to everything which can be called writing."

Being an obedient fellow, young Semmelweis enrolled as a law student at the University of Vienna in the autumn of 1837, when he was nineteen years old. A legal degree was the preferred choice of many bourgeois families for their sons, because it opened the door to a variety of opportunities in the civil service and other areas of public life. His father wanted him to become a military judge, but the law proved to have little appeal. One day, after accompanying a medical student friend to an anatomy lecture by the charismatic professor Joseph Berres, Ignác resolved to change the direction of his career. By the following year, he was enrolled in the medical school.

After a year of studying medicine in Vienna, Semmelweis continued at the University of Pest, where he stayed for two years before returning in the autumn of 1841 to complete the qualifications for his degree. Not only was the University of Vienna a far better school, but its graduates had the privilege of practicing anywhere they might wish in the entire Austrian Empire, whereas those of Pest were restricted to Hungary. Accordingly, such a back-and-forth migration was a relatively common route to an academic degree. It provided a way to study close to home for as long as possible, and also the means of obtaining an unrestricted diploma. By February 1844, Semmelweis had fulfilled all the requirements, including the writing of a dissertation called *Tractatus de Vita Plantarum*—in the field of natural history. He was scheduled to receive his degree in April, but his mother's unexpected death made him hurry back to Buda, postponing his official graduation for some six weeks. He signed a statement in the faculty register affirming that he intended to return to his own country.

That country was a restless land. Hungary was part of the vast Austrian Empire, a polyglot agglomeration of peoples whose major groups were the Germans, Poles, Hungarians, and Italians, and among whose lesser were Czechs, Slovaks, Ruthenians, Serbs, Croats, Slovenes, Dalmatians, and Romanians. The entire assortment of nationalities was under the domination of the Habsburg emperors, ruling from Vienna through the repressive influence of their first minister, Prince Klemens von Metternich. Fearful of the rising tide of cultural nationalism and accompanying stirrings of political agitation—particularly among Germans, Poles, Italians, and Hungarians—Metternich established policies designed to quash attempts at liberal reform and expressions of autonomy. His strategies of forcing Germanic culture on the various countries and playing off one national cluster against the other were particularly troublesome in Hungary, where lesser ethnic groups were seething under the dominant Magyars. To the leaders of the empire, Hungary was a backward province, requiring the example of Austria to lead it toward modernity.

But discontent was rife in Austria, too, particularly among the students at the University of Vienna. Though Hungarians formed the largest group of foreigners, much of the rest of the student body also came from the various lands of the empire. Not only they, but the Austrian students and younger faculty members as well, chafed under the stifling control of government ministries and conservative older professors. By the eve of the uprisings of 1848, the university, and most particularly its medical school, had become a hotbed of revolutionary activity in both the political and the educational sense.

Writing some thirty years later, the great German surgeon Theodor Billroth, the author of the most comprehensive study of medical education until that time undertaken in Europe (*The Medical Sciences in the German Universities,* 1876), would describe the older faculty of the University of Vienna as

[a] generation that had been reared in an intellectual strait-jacket with dark spectacles before their eyes and cot-

ton wool in their ears. The young people turned somer-
saults in the grass, and the old men, whose bodies had been
hindered in their natural development by the lifelong burden of state supervision, felt their world tumbling about
their ears, and believed that the end of things was at hand.

Those were heady days. When, on March 12, 1848, the students and some of the younger faculty made their revolutionary demands, it was under the slogan of *Lehrfreiheit und
Lernfreiheit,* "freedom of teaching and freedom of learning."
When similar protests occurred on campuses across America
and Europe in the late 1960s, their leaders, very likely unbe-
knownst to themselves, derived their rallying cry of "Academic
freedom!" directly from that place and that time.

It was into the early phases of this ferment of emotionally
charged activity that Ignác Semmelweis had thrust himself
when he entered medical school. Having become fascinated
by the research being done in pathological anatomy, he
applied upon graduation for an assistantship with Jakob
Kolletschka, a Rokitansky disciple who was using his mentor's principles in the study of forensic pathology. Because
Kolletschka was a young investigator much admired by the
newly minted physician, the latter's disappointment was great
when his application was denied.

But Semmelweis did not allow his hopes to remain dashed
for long; he next applied to be assistant to Joseph Skoda, the
leading physician of the medical school. Skoda was doing
important work on percussion and auscultation, in which,
like Rokitansky, he correlated clinical findings with the
pathological changes that caused them. But the professor had
promised the job to Gustav Loebl, a recent graduate who was

Semmelweis's senior, so another rejection had to be dealt with. Semmelweis turned his attention to obstetrics.

Obstetrics did not occupy an elevated position in the academic hierarchy of mid-nineteenth-century European medicine. Though chairs in the discipline had been established at all major universities, the subject remained an elective course for most students. The great majority of deliveries were being performed by midwives, especially the many done outside of hospitals. One cannot help wondering whether Semmelweis's quite natural apprehension about the possibility of yet another rejection caused him to play it safe by choosing a specialty in which a training position was easier to obtain.

The history of the obstetrics department of the Vienna General Hospital provides a preview of the events that would later overtake it. To understand its development, it is necessary to review the history of the hospital itself and to turn to the fifteen-year reign of Empress Maria Theresa, who ruled jointly with her son Joseph II from the death of her husband, the Holy Roman Emperor Francis I, in 1765, until her death in 1780. Guided by Dr. Gerard van Swieten, one of the most far-thinking physicians of the eighteenth century, she instituted numerous humanitarian reforms in public health and medicine. Toward the end of her life, she laid plans for a great hospital in the capital city, which her son Joseph II pursued with the same enthusiasm that he devoted to furthering the development of the University of Vienna. To strengthen its academic programs, he sought out leading scientists and physicians of the time, and gave financial and moral support to the medical school. In 1784, he fulfilled his mother's great vision by establishing the Allgemeine Krankenhaus, including among its buildings the largest lying-in hospital in the world.

Pursuing the policies and commitments of his mother, Joseph paid particular attention to obstetrics. Determined to make that department the best in Europe, he chose the thirty-three-year-old Dr. Johann Boër to be its head and sent him to study in France and England. After a year working in London, Boër returned to Vienna and was named professor of midwifery and director of the lying-in hospital. Much impressed with the gentleness of technique and the natural methods he observed in the practices of several of the British accoucheurs, he instituted similar procedures at the Allgemeine Krankenhaus. He ordered his staff to limit the number of internal examinations during labor and to turn to instrumental delivery only as a last resort, making it clear that he would stand by the dictum that he articulated in the statement. "Whatever is done artificially, apart from this given necessity, is mere bungling, serving only to torture and destroy mother and child." Boër put his faith in the healing powers of nature; in his eyes, the emphasis placed on pathological anatomy was unwarranted and would only lead to meddling.

Boër's respect for the humanity of parturient women extended to those who died of puerperal fever or other complications. He dissected their bodies only to study the pathologies that had killed them, and refused to allow instruction on the corpses of dead mothers. This was a departure from the practice prescribed in the formal curriculum. Instead, he used a painted wood model, popularly known as "the phantom," to teach pelvic anatomy.

The result of all of these measures was a profound decrease in the incidence of childbed fever. With the exception of only a few brief periods, its mortality was consistently around 1 percent during the more than thirty years of his tenure. All of this changed when his successor, Johann Klein, took over clinical responsibilities in 1823, after Boër's final year, when the death rate was 0.84 percent. Klein, being a man beholden to political influence for his appointment, was less willing to defy the strict requirements of the curriculum, or to vary his methods from those generally accepted in obstetric practice. He reinstituted the requirement that students learn to do obstetric examinations by using cadavers, and he relaxed his predecessor's constraints on internal examinations during labor and forceps deliveries.

There were plenty of cadavers to work on. Government regulations required that an autopsy be done on every patient who died while hospitalized. Actually, the practice of performing autopsies on women who died in lying-in hospitals was the rule in virtually all European institutions, with the exception of those in Great Britain and Ireland.

Following Klein's ascension, the mortality rate immediately rose to 7.45 percent. With periodic fluctuations, it would remain at that level as long as he was in charge. A second maternity division was established under his control in 1834, being used solely for the training of midwives after 1839. Though its mortality figures were always about a third those of the First Division, where only medical students and obstetricians practiced, no lessons seemed to have been learned by the director. He continued to insist that the deaths were due to "epidemic influences." This was a man who epitomized those professors who "had been reared in an intellectual strait-jacket with dark spectacles before their eyes and cotton wool in their ears."

It was in Klein's interest to remain that way. The Austrian

Empire being then dominated by Metternich's personality and policies, control of major faculty positions in the university was in the hands of government ministers, who made sure that their appointees hewed to the official line in all things. Policies were conservative, opposition was stifled, and every professor knew that maintaining his post was contingent on maintaining the status quo. Johann Klein was among the staunchest of the old guard and a good friend of the politicians in the health ministry. When Ignác Semmelweis applied to be his assistant, the fifty-six-year-old chief had been director for more than two decades and become increasingly resistant to any thought of change. In his treatment of patients and in his methods of ruling over his staff, Klein was the sort of man whom Boër would have dismissed as incompetent and autocratic.

Prior to applying for Klein's program, Semmelweis had to obtain the diploma of master of midwifery. In his usual thorough manner, he took the preparatory course work twice and passed the examination in August 1844, having presented himself as a candidate for the post of assistant one month earlier. He received notice of his appointment in February 1846, and took over the post on July 1.

Semmelweis did not remain idle during the two-year interval between his successful examination and the day he finally took on the duties of assistant to Klein. Not only did he have free access to the clinic, but Rokitansky generously allowed him to dissect the bodies of women who had died of gynecological diseases and operations. The professor encouraged him in his work and saw to it that he learned a great deal about the new methods of observation and analysis. When he at last went to work on the wards, it was with the perspective of an experienced pathological anatomist. But he had to give up the post four months later because his predecessor, Dr. Breit, elected to extend his term.

Semmelweis used his period of enforced freedom to resume his determined efforts to prepare himself for a thorough understanding of the diseases of women. Profoundly affected by his experience with puerperal fever during the preceding four months on the wards, he undertook the study of English so that he might spend a period working at the renowned Dublin Lying-in Hospital, in the hope of learning some new things about the disease far from the stifling political atmosphere of the Allgemeine Krankenhaus. But his plans were changed when Breit, at the end of February 1847, was appointed professor of midwifery at the University of Tübingen. With two friends, Semmelweis took a brief holiday in Venice before plunging himself once more into his duties in the First Division. On March 20, 1847, he finally ascended to a two-year appointment as assistant in obstetrics.

There was no reason for Klein to believe that his twenty-eight-year-old assistant was anything other than the good-natured, well-liked young man he appeared to be. Of middle height and somewhat stout, Semmelweis had already begun to lose his blond hair. His gray-blue eyes smiled easily when colleagues made fun of his Buda-Swabian accent, and he seemed not to mind being the butt of occasional jokes at his own expense. Contemporary accounts refer to his "playful, jocular nature" and describe him as "lighthearted" and "popular." There was as yet no evidence of certain unpleasant traits that would emerge in later years, described by his first English biographer, William Sinclair, as "the most explosive indignation and sarcastic contempt for fools" who could not grasp the truth of his discovery.

Once he took over the assistantship, Semmelweis devoted all of his energies to his work on the wards and his continuing dissections, focused, as they had been during his earlier four months, on the bodies of women dead of puerperal fever. There were so many cases of the disease in the First Division and on the morgue slabs that physicians from various parts of Europe would visit in order to study them. The ringing of the pastor's bell was a familiar sound, as priests entered the building that housed the obstetric units to administer the sacraments, at all hours of the day and night, to women awaiting death in the rooms to which the sickest were taken. Wearing his ceremonial robes, a priest would walk through the ward preceded by his assistant waving the small handheld bell, which had the effect of frightening the many parturient women who heard it. This, too, became a favorite cause of childbed fever in the minds of some naïve theorists who believed that fear caused the disease. Semmelweis himself came to listen for the sound with trepidation. Reflecting on those episodes in later years, he wrote, "Even to me myself it had a strange effect upon my nerves when I heard the bell hurried past my door; a sigh would escape my heart for the victim that once more was claimed by an unknown power. The bell was a painful exhortation to me to search for this unknown cause with all of my might. Even to this difference between the arrangements of the two Divisions was attributed the higher mortality of the First Division." During the brief four months of his first aborted assistantship, he appealed to the humanity of the clergymen, begging them to take a roundabout way to the death rooms. Apparently, no one had thought to do this before, and it seems never to have entered the minds of the priests themselves. As Semmelweis expected, the changed route had no effect on the prevalence of disease.

But to those who believed that fear was a significant instigating element in the causation of puerperal fever at the Allgemeine Krankenhaus, the ringing of the bell was only one of several possible contributing factors. Chief among them was fear of the institution itself, and specifically of the First Division. Semmelweis had ample opportunity to observe its manifestations during his first brief term as assistant:

> That they were afraid of the First Division there was abundant evidence. Many heart-rending scenes occurred when patients found out that they had entered the First Division by mistake. They knelt down, wrung their hands and begged that they might be discharged. Lying-in patients with uncountable pulse, meteoric abdomen [the rapid onset of distension], and dry tongue, only a few hours before their death, would protest that they were really quite well, in order to avoid medical treatment, for they believed that the doctor's interference was always the precursor of death.

Those who espoused such theories clung to them despite the obvious fact that the dread was the result of the prevalence of disease, and not the other way around. Equally fanciful notions were bandied about in the hospital, but the silliest of them must surely have been the conviction by some that the mere presence of male accoucheurs wounded the modesty of parturient mothers, leading to the pathological changes that took their lives. This fantasy was meant to explain the much higher incidence of disease in the First Division. It could not account for the fact that male obstetricians almost universally attended mothers of the upper classes of society at their home deliveries, where cases of puerperal fever were rare. Such women could hardly be thought less modest than the occupants of beds on the public wards of the Allgemeine Krankenhaus.

Other explanations were freely offered as well, but all of them seemed even at the time to be grasping at straws, without any evidence to support them. Catching a chill, errors in diet, rising in the labor room too soon after delivery in order to walk back to bed—the hospital was rife with such speculations. None of them took account of the observable facts, and most applied equally to both divisions. But to the major theorists of the institution, specifically Johann Klein and those in his orbit, puerperal fever was the result of some unspecified influence that hovered over the First Division more than over the Second. It was a *genius epidemicus* that was the culprit, they claimed—some specific variety of epidemic against which they were not only powerless but blameless as well.

Semmelweis observed all of these things during those first four months, and even at that early time in his career he knew that none of the explanations being offered were tenable. To admit to the beds of the First Division healthy women of his own age and younger—blooming with the wholesome flush of imminent childbirth—and then to see them sicken and die before his unbelieving eyes within days was his tragic initiation into the fraternity of obstetricians. Perhaps even worse was to bear witness to the stinking devastation of putrefying flesh that he observed upon opening their bodies in the dissecting room only hours later. Surely, he could not have been the only young doctor to be horrified by what he saw on the wards and in the autopsy room; surely, he could not have been the only young doctor to conjure with the disquieting thought that in some way he had himself contributed to the ruin of individuals and of families; but just as surely, he *was* the only young doctor who convinced himself that a way could be found to end the carnage, and that he would be the one to do it.

V

The most challenging question facing the newly appointed assistant was the nature of the disease itself. What, in fact, was puerperal fever? He quite correctly realized that, for all the windy pontifications he had heard and read, no one really understood it as an entire pathological entity. Various authors had placed differing emphasis on one or another of the findings, depending on the abnormalities with which they were most impressed. Was it a form of peritonitis? Was it a disease of the uterus? Of the tubes? Of the omentum? Was it a phlebitis? Was there some indeterminate degeneration of the blood? In this period when dissectors were doing their best to relate symptoms to the organs in which they originated, specificity was all the rage. As though in reaction to the centuries in which generalized diseases caused by such universal etiologies as humoral imbalance had dominated medical thought, the Morgagni-inspired anatomical approach aimed at localizing everything it encountered. But as so often occurs when old theories are discarded for new, the pendulum sometimes swung too far. This was precisely the case in the attempt to understand puerperal fever.

The studies Semmelweis had done while waiting for his assistantship did not end when he finally attained it. In addition to carrying out his other duties, he made sure to dissect cadavers every morning before going to the wards for his clinical work. The students and young doctors in training were also there, either doing their own dissections or using the bodies to practice internal examination as the curriculum required. In his investigations, Semmelweis was guided by the principles established through the work of the French physicians after the turn of the nineteenth century: careful physical examination followed by autopsy after the patient's death, to identify the suffering organs from which the cries originated. Material with which to study puerperal fever was available in profusion—during some months, as many as 30 percent of postpartum mothers died of the disease.

[F]rom the day I decided to devote my life to midwifery, i.e from 1844 to my removal to Pest in 1850, I was accustomed, before the morning visit of the Professor, to examine for the benefit of my gynecological studies almost every day all the female bodies in the deadhouse of the Imperial and Royal General Hospital. The kindness of Professor Rokitansky, of whose friendship I could boast, gave the opportunity to dissect all the female cadavers, which were not already set aside for autopsy, so that I could check the results of my [physical] examinations by the results of the dissections.

And thanks also to Rokitansky, Semmelweis had become as skilled an observer as he was a dissector. Though he, too, was concerned with tracking symptoms to their individual origins, he appreciated the greater picture, namely that comprehensive knowledge of disease can be achieved only by the evaluation of resemblances between its effects in the various organs and tissues. Having identified those pathological manifestations in differing structures, one attained ultimate understanding by seeking out relationships between them. He had learned from his great teacher that the identification of symptoms with their organs of origin is not enough—the greatest lesson of pathological anatomy is that *patterns* of findings are as important as the findings themselves. As a perceptive student of his mentor, he was unlikely to confuse trees with forest. Where others saw differences, he looked for similarities. With respect to puerperal fever, he was convinced that there had to be some all-embracing pathology that united everything he observed in the many corpses that he was tirelessly dissecting.

To comprehend the nature of puerperal fever was the theoretical conundrum, but the *pragmatic* challenge was of far greater urgency: in 1846, the year before Semmelweis ascended to the assistantship after Breit finally left, 459 women had died of the disease in the First Division. Because this number did not include those patients transferred to the internal medicine or other divisions when they became very sick, the true figure was doubtless considerably higher. As always when bad statistics were piling up, a great deal of discussion had ensued during the year, minor modifications in procedures, instruments, and teaching methods had been instituted, and much hand-wringing was displayed at all levels of the medical staff—and yet another investigating committee composed mostly of nonmedical functionaries was appointed by the government, in the same useless way as in several of the preceding years. In the face of its conclusion—and Klein's self-exculpating theory—that a *genius epidemicus* hung over the city, little was made of the fact that only one-fourth as many mothers, 105, had succumbed to the disease in the Second Division. So long as such a miasma was blamed, the wringing hands could consider themselves also tied. As far as could be seen, there was no solution to be found.

Not only epidemic was being blamed—Klein taught the milk theory to explain the method by which the epidemic achieved its objective. To his credit and that of the commission, its members did note that many medical students examined each woman in labor, and it was therefore recommended that the number of foreign students be greatly decreased. This resulted in a temporary reduction in mortality, but it returned to its original levels in the first months of 1847. The attempt to shift blame to students, and the foreign ones in particular, did not achieve its objective. But at least some acknowledgment was made that there might be a connection between the amount of medical intrusiveness to which a patient was subjected and the probability of subsequent sickness.

Klein was willing to grasp at any observable characteristic of the First Division that would explain why its patients were dying at a higher rate than those of the Second, so long as it left him and his colleagues without fault. At the meeting of the 1846 commission, for example, he brought up the poor condition of the division's walls as a cause to be considered.

This was more than Semmelweis, then in the five-month period between his first and second appointments, could bear. Though a junior physician should certainly not have been so undiplomatic, he rose to point out that many maternity hospitals had walls in far worse condition, and yet their mortality rates did not approach that of the First Division. This was hardly the sort of behavior that would endear him to the professor.

In fact, the relationship between the two men was already becoming uneasy. Semmelweis had made no secret of his dissatisfaction with all the current explanations of puerperal fever. Moreover, it was clear that he was determined to find the solution to the problem of its etiology and prevention. Such behavior was not calculated to sit well with Klein. As far as he was concerned, the mystery had been solved and the disease was beyond prevention or cure by the doctors; this was the way he wanted to see it. He was not a man without a conscience, but, like so many other obstetricians of the time, he needed to placate it by believing himself to be helpless in the grip of forces beyond his control. And he needed also to keep his job.

Even Semmelweis found himself turning toward small modifications in technique in the hope of lessening the ravages of the disease. Though his aim continued to be an all-out search for causes and for methods of prevention, he instituted technical variations based more on optimism than on reality. Because the midwives delivered women lying on their sides, he began to do the same; he had mothers carried back to the ward after delivery instead of the usual practice of making them walk, in a theoretical attempt to decrease trauma; he changed the way drugs were administered; he saw to it that ventilation of the ward was adequate—all to no avail. He was, in his own words, "like a drowning man, who grasps at a straw." But nothing seemed to change the incidence of disease. Moreover, the young obstetrician's state of mind was not eased by his now chilled relationship with the chief. Nor was it improved by the death of his father during the interim between the two phases of his assistantship. Semmelweis's vacation in Venice, in fact, was suggested by one of his companions, Lajos Markusovszky, because his depressed spirits needed refreshing. There was much to think about, and all of it was sad—his father's death, Klein's quite obvious dislike, and the carnage that he had long been witnessing among the young women who came to the First Division.

One should not claim for Semmelweis that he was more sensitive to the plight of the dying mothers than were the doctors working alongside him. But, somehow, neither the notion of intellectual challenge nor the suggestion that he was motivated by driving ambition can suffice to explain the single-minded campaign that the twenty-eight-year-old obstetrician commenced against the disease as soon as he assumed the duties of assistant. It was as if he had convinced himself that the full responsibility for victory over it lay on his bulky shoulders.

Those seeking psychological factors that might have given Semmelweis his herculean zeal would be justified in pointing out that his mother had died only one week before he was scheduled to receive his medical degree. Though making much of this fact would be speculative at best, it should not be discarded as a possible source of the wellsprings of emotion that seem to have driven him toward his goal. No description has been left of his overt reaction to his mother's death, but it hardly needs to be said that such life-altering events are not without major consequences in one's conscious and unconscious mind. At such an instant in a man's life, to begin the dedicated dissection of the bodies of dead mothers in an attempt to prevent further deaths should be treated as suggestive evidence well beyond the merely circumstantial or coincidental.

In this regard, the descriptions of his personality from the period of the great discovery differ markedly from those that precede his graduation from medical school in 1844 or the beginning of his assistantship. Terms like "bright and jolly companion" and "playful, jocular nature" have been used in two of the major Semmelweis biographies, those by Sinclair and by Gortvay and Zoltán, respectively, to describe him as he was before completing his degree requirements at the University of Vienna, and there is no disagreement in other sources. But later, around the time of his discovery, a significant change clearly occurred, to a brooding intensity that was remarkably different from the temperament of the "light-hearted," "smiling," and "popular"—to use a few of the adjectives invoked by these authors and others—young man he had previously been. It is hard to escape the conclusion that his mother's death, the changes in his personality, and the commencement of his unremitting search for the etiology of childbed fever are related in more than the temporal sense. Another crushing blow was surely added with the death of József Semmelweis just before his son began the second period of his assistantship.

With the beginning of the abruptly cut-off first period, Semmelweis's studies of puerperal fever increased in volume and determination. He became personally responsible for the welfare of the women assigned to the First Division, and the sense of urgency intensified. He sought knowledge in the library, the autopsy room, and at the bedside, and few of his waking hours were spent elsewhere. He would have to digest a great deal of information before he could make any sense of the findings he was accumulating by dissection and the evaluation of patients and cadavers.

There was much to study in books and journals. A huge mass of writings had been published over the decades, and it was necessary to peruse all of it, lest some crucial insight be overlooked. Even more than before, Semmelweis familiarized himself with the opinions of the English contagionists and their opponents. He studied the evidence—or lack of it— upon which the various theories of etiology had been founded, whether cosmo-telluric or based on interpretation of autopsies. As he evaluated his own patients and dissected more and more bodies, it became increasingly clear that his gradually evolving perception was accurate: there was little or no correlation between the several main theories to which the leaders of European obstetrics subscribed and the actual evidence as it was accumulating before his eyes.

Among the most critical of the qualities characterizing the successful researcher is the ability to separate wheat from chaff. From the wealth of data, observations, and background information that face a scrupulous investigator, he must be able to extract those that have real significance, before he can proceed to build an edifice upon which to base his under-standing of the problem he is trying to solve. And here, Semmelweis was unerring in his accuracy. He filtered out the inconsequential and came up with the most important of the observations that could be verified by anyone objectively viewing the conditions then obtaining at the Allgemeine Krankenhaus. During his waiting periods and in his brief first time as assistant, he had become convinced that several of them were critical to the evaluation of evidence.

Observation no. 1: The same number of deliveries took place in each of the hospital's two obstetrical divisions, usually between 3,000 and 3,500. The only difference between them was that deliveries in the First Division were carried out by doctors and medical students and those in the Second Division by midwives and students of midwifery. In the First Division, an average of 600 to 800 mothers died each year from puerperal fever; in the Second Division, the figure was usually about 60 deaths, one-tenth as many.

Observation no. 2: Although childbed fever raged violently in the First Division, there was no such epidemic outside the hospital walls, in the city of Vienna. The mortality of home delivery, whether by midwives or private doctors, was low. Even when they gave birth in alleyways and streets, the so-called *Gassengeburten*, mothers who self-delivered rarely died.

Observation no. 3: Despite a general impression to the contrary, the decades of carefully kept statistics at the Allgemeine Krankenhaus showed that neither the incidence nor the mortality of puerperal fever was related to weather, as epidemics often were.

Observation no. 4: Greater degrees of trauma during delivery appeared to increase the likelihood that a mother would develop puerperal fever. This was true of no other epidemic disease.

Observation no. 5: Closing down the ward for a period would always stop the mortality. When mothers were delivered elsewhere during that time, they did not get sick.

Observation no. 6: The infant delivered of a woman who subsequently died of puerperal fever would itself not infrequently die of a fever similar to its mother's. In such cases, the findings on autopsy were similar to those identified in the mother.

When Semmelweis returned from Venice in 1847 to take up his duties, he was already convinced of the critical importance of these easily demonstrable facts, by which it had become clear to him that neither epidemic nor miasma could be the cause of puerperal fever. If he were to identify its ultimate etiology, he would have to seek it in some factor endemic in the First Division and in the First Division only. Also, he was coming to the conclusion that the process was transmitted by direct contact between individuals, because the newborns quite obviously were dying of the same disease that had killed their mothers. He returned to work determined that he was gradually closing in on his goal. But if the respite from patient care had helped raise his spirits at all, they were dashed again by the news he received on his arrival back in Vienna on March 20, 1847. Jakob Kolletschka, so admired by Semmelweis not only because of his warm personality but for his adherence to Rokitansky's principles in forensic pathology, was dead. He had been accidentally stuck in the finger by a student's knife while conducting a medicolegal autopsy, and died of massive infection within a few days. When his body was dissected, its organs and tissues were permeated with pus and abnormalities exactly like those common in women who succumbed to puerperal fever, and sometimes in their infants.

Semmelweis's description of the effect these findings had

on him, as he reviewed his friend's autopsy transcript again and again, is striking. Though he was a notoriously inept and verbose writer whose narrative frequently goes astray in convoluted prose and ill-natured diatribe, he is in this one instance crystal clear, even amid some of his usual clumsiness of language. I have chosen here to use the text of his words as they are to be found in the translation done in 1968 by Ferenc Gyorgyey (later the director of the Yale Medical Historical Library) in the course of writing his master's thesis in the history of science at Yale University. The Gyorgyey translation was made from Semmelweis's lectures in Budapest, published in Hungarian in the medical journal *Orvosi Hetilap* in 1858, three years before he brought his thesis to a much wider audience, in what was then virtually the universal language of medicine, as *Die Aetiologie, der Begriff und die Prophylaxis des Kindbettfiebers* (The Etiology, the Concept, and the Prophylaxis of Childbed Fever). I use the Gyorgyey version because of its clarity, as opposed to the verbal ineptness that would be encountered in reading a translation from the German of the *Aetiologie*. There can be no more dramatic description of the moment of great discovery.

Totally shattered, I brooded over the case with intense emotion until suddenly a thought crossed my mind; at once it became clear to me that childbed fever, the fatal sickness of the newborn and the disease of Professor Kolletschka were one and the same, because they all consist pathologically of the same anatomic changes. If, therefore, in the case of Professor Kolletschka general sepsis [contamination of the blood] arose from the inoculation of cadaver particles, then puerperal fever must originate from the same source. Now it was only necessary to decide from where and by what means the putrid cadaver particles were introduced into the delivery cases. The fact of the matter is that the transmitting source of those cadaver particles was to be found in the hands of the students and attending physicians.

The source of the cadaver particles seemed obvious: they were being transmitted by the hands of the students and attending physicians who came to the First Division directly from their labors in the autopsy room, where they had been examining the bodies of women freshly dead of puerperal fever. Because there seemed no reason for them to wash their hands, except superficially, or change clothing before coming to the First Division, they did neither. They brought the invisible matter directly from its source to the women in labor. Once introduced onto the external genitalia and inoculated into the vagina and cervix, the particles were absorbed into the bloodstream and lymphatic channels, being by this means transported to other tissues and organs. Transmission might also occur to the soon-to-be-born fetus, because its vessels and those of the mother were still in communication through the placenta. In addition, the immediate postpartum mother presented a particularly favorable site for absorption via the denuded area where the placenta had been. Greater degrees of trauma resulted in greater numbers of open channels by which the particles might enter the mother's circulation. In the words of Semmelweis, "puerperal fever was nothing more or less than cadaveric blood poisoning."

In a single insight, Semmelweis had solved both parts of the problem he had set for himself—to identify the nature of puerperal fever and to find a way to prevent it. He now knew that the disease was a contamination of the blood by cadaver particles; its means of transmission was the hands of students and doctors; it could be prevented by ridding them of the infecting material; the presence of cadaver particles was recognizable by the odor it imparted to everything on which it adhered.

Chloride solutions had long been used to rid objects of the noxious odor of putrid materials. Semmelweis reasoned that a chloride solution would be the ideal substance to destroy the foul-smelling cadaver particles. In the middle of May 1847, he ordered that a bowl of *chlorina liquida*, a dilute concentration of the disinfectant, be placed at the entrance to the First Division, and he insisted that every entering medical attendant wash in it before touching a woman in labor. Small stiff brushes were kept nearby, to be used for cleaning under fingernails. After a short time, chloride of lime was substituted because it was less expensive.

The salubrious results of the new prophylaxis took a few weeks to become apparent. But by early June there was no question that something remarkable had begun to happen. During the last seven months of the year, only 56 women died, of the 1,841 delivered in the First Division. This meant that the mortality had fallen to 3 percent, comparable to that of the Second Division. In 1848, the first full year of the program, the First Division had a puerperal death rate of 1.2 percent and the Second Division of 1.3 percent, virtually equal. In March and August of that year, not a single death occurred in the First Division. Other than the practice of chlorine hand cleansing, only one change had been made. Klein convinced himself that a recently installed ventilation system was the reason for the startlingly improved statistics. It was the only way he could rationalize the facts staring him in the face, and he clung to it as to an amulet.

But to Semmelweis, as to everyone else who looked at his thesis and results objectively, the striking difference between the mortality in the two divisions was easily explainable. Since neither the midwifery students nor their teachers had occasion to handle cadavers, they were not contaminating themselves.

And further support for the new theory was yet to come. In October of 1847, a woman in labor in the First Division was discovered to have an advanced cancer of the breast, which had ulcerated and was draining foul-smelling pus. Eleven of the twelve other women on the ward died of puerperal fever. In the following month, a laboring mother with an infected knee joint was admitted, with the result that yet another group of patients succumbed. At first, Semmelweis believed that the wound of this second woman had contaminated the atmosphere of the ward, but in time it became obvious that the hands of nurses dressing her knee had been the means of transmitting infectious material. It was becoming clear to him that whether the invisible particles came from the cadaver or some other source, the cause of the disease was absorption of decomposed organic matter.

All of these events were consistent with certain facts in the history of the lying-in institution. They are worth reviewing at this point in the narrative. Upon becoming its director in 1789, five years after it opened, Boër had introduced the methods he learned from his period spent studying with English obstetricians. The hallmarks of their approach were cleanliness, gentleness, and patience. In addition, he refused to use cadavers in the teaching of students, employing an anatomical model instead. Of the 65,000 patients delivered during his thirty-three-year tenure, only 1.3 percent died, whether of puerperal fever or any other cause; in his final year, the mortality was 0.84 percent. It was now completely comprehensible that Klein, who taught on cadavers and used far less gentle techniques during labor and delivery, should have had a high mortality during the entire period of his directorship. Moreover, it had long been known that the incidence of puerperal fever declined when the amount of teaching declined. That all of this information was readily available was due to the careful record keeping that had characterized the Allgemeine Krankenhaus since its beginnings. In few other European or American institutions was such reliable documentation available.

One of the remarkable phenomena familiar to the entire staff of the First Division was the frequency with which mothers would become sick in rows all at one time. The Semmelweis theory could now elucidate the reason for this phenomenon, as follows. No matter the other probings they might undergo, each woman in labor was formally examined on rounds twice each day, in the morning by the professor and in the afternoon by the assistant. They would proceed from one bed to the next, contaminating each patient in sequence. After chloride of lime hand washing was instituted, there was never another outbreak in a row of puerperae.

And then there were observations that Semmelweis himself had made, which now came into perspective. He would grapple with one of them for the rest of his life: he had noted over the three years that whenever Dr. Breit was less than attentive to his duty to dissect each morning, the death rate fell. Conversely, should a staff member take a special interest in some aspect of pathological anatomy, the mortality in the ward would rise during the period when he was particularly active in the dissecting room. Of no one was this more true than Semmelweis himself:

> Because of my convictions, I must here confess that God only knows the number of patients who have gone to their graves prematurely by my fault. I have handled cadavers extensively, more than most accoucheurs. If I say the same of another physician, it is only to bring to light a truth, which was unknown for many centuries with direful results for the human race. As painful and depressing, indeed, as such an acknowledgment is, still the remedy does not lie in concealment and this misfortune should not persist forever, for the truth must be made known to all concerned.

The principles of what would become known as the Semmelweis *Lehre,* or doctrine, were now established. They were simple to understand, logical, and consistent with every observation the young obstetrician had made during the three years of his quest. Moreover, they were a direct outgrowth of the thinking patterns, investigative techniques, and reasoning methods he had learned from Rokitansky. In every way, Semmelweis's accomplishment epitomized the new methods introduced by the chief of surgical pathology, and stood in opposition to the reactionary complacency of the old guard.

The two camps were separated by the wide gulf existing between the burgeoning investigative science and the fuzzy theoretics that had characterized medical thought for centuries. And they were also separated by age. Rokitansky, the exemplar of the new, was forty-three at the time of the Semmelweis discovery, and had been director of pathological anatomy for three exhilarating years; Klein, the exemplar of the old, was almost sixty, and had headed the maternity clinic of the Allgemeine Krankenhaus for twenty-five stifling years.

And it is hardly necessary to add that the two men were separated as well by politics. For his appointment and for maintaining his position, Klein was obligated to powerful government bureaucrats whom he had at all times to satisfy. Rokitansky, on the other hand, had no political debts. He owed his place and his stature in the university solely to his enormous ability and the contributions he was making to medical science. At the time of Semmelweis's elevation to assistant, the groups represented by the two professors were in open conflict. The battle lines had already been drawn.

Into this setting—the new versus the old; the intellectually liberal versus the conservative; the scientific study of disease processes versus staid adherence to increasingly discredited notions—the Semmelweis *Lehre* figuratively exploded. The entrenched older faculty resented the younger men, whose liberal policies and new ideas in research and medical theory they opposed. The younger men chafed under the stifling control of their superannuated seniors and the government ministries to which they were in thrall. Moreover, Europe stood on the brink of the revolutions of 1848. In the political sense, those uprisings would fail. But in the academic arena certain real gains would be made. As though straining toward that hope, the emerging young leaders of the new medicine were preparing themselves for the coming confrontation. There could not have been a better field on which to engage in combat than the puerperal fever theory of Ignác Semmelweis. It was the embodiment of the philosophy of the young faculty; it was the embodiment of everything the old faculty was resisting. There is some poetic truth in saying that what Semmelweis would, years later, call "the rising of the Puerperal Sun over Vienna" was to signal the call to arms for this major battle in the history of the Vienna Medical School.

The battle would eventually be won by the younger faculty, and the most glorious period in the long saga of Vienna medicine was ushered in by the victory. In less than two decades from this time, the Allgemeine Krankenhaus and the medical school of which it was a part became the most popular of the clinics visited by European and American physicians seeking to learn the latest in research, bedside medicine, operative surgery, and teaching methods. They came not only to visit but to take classes and to work in laboratories. It has been estimated that some ten thousand Americans, for example, engaged in periods of formal medical study in Vienna between 1870 and 1914. And all of this began to happen as a result of the campaign that was being framed by the activities of the young faculty members of the medical school, beginning in those heady days of the late 1840s.

Rokitansky was the most prominent member of this relatively small group, and the mentor of the rest. But there were two others—each of them trained in the master's methods and philosophies—whose work complemented his in forging the future.

The first of these was Joseph Skoda, the outstanding clinician of the Vienna Medical School. Since the invention of the stethoscope in 1816, many leading physicians had studied the sounds made audible by its conductive capacities, and attempted to identify ways in which they could be used in the diagnosis of disease. Much the same was true of percussion, the method of tapping on the chest in order to ascertain the presence of pathological changes in the lungs, heart, and thoracic cavity. But no other physician was as successful as Skoda. Using Rokitansky's methods, he made scrupulous observations of the autopsy findings associated with the sounds he had been able to elicit before death, and became so accurate in differentiating one tone from another that he could detect and categorize subtle dissimilarities that, properly interpreted, distinguished individual pathological processes and clinical entities. The precision of his methods lent itself to teaching them to others, with the result that Skoda became renowned throughout the medical world as the founder of a tradition in physical diagnosis. Not only that, but his methods were so accurate that they made possible virtually immediate diagnosis of many conditions, giving rise to the reputation of the Vienna Medical School as the leading proponent of what became known as *Schnell-Diagnosen*, diagnoses that were lightning quick.

Unlike the kindly and generous Rokitansky, Skoda was a man whose heart seemed inaccessible to personal warmth. He was well known to be cold and distant, which seemed of a piece with his approach to medicine and patients. His interest was in the scientific basis of diagnostic methods, and he valued the detachment that enabled him to achieve the objectivity he considered crucial to accuracy. Once having identified a disease, he cared not a whit for therapy, becoming well known among his colleagues for shrugging his shoulders when consulted about such matters, as though to brush off the importance of the inquiry. He would then bark out his standard reply, *"Ach, da ist ja alles eins!"* (They're all the same!). Not suprisingly, the emotionally remote Skoda remained a lifelong bachelor.

Skoda's method of clinical reasoning had caught the imagination of Semmelweis during his medical school years. It was for this reason that he had applied, albeit unsuccessfully, to be the master clinician's assistant. But Skoda had kept a close eye on the younger man's progress in applying his and Rokitansky's methods to the riddle of puerperal fever. Both professors realized that Semmelweis's discovery was a logical outcome of their own teachings in scientific logic. The historian Erna Lesky, in her encyclopedic history entitled *The Vienna Medical School of the 19th Century,* states that these two rising giants were the "intellectual fathers of his discovery."

The second exemplar of the changing philosophies of disease was Ferdinand Hebra. Hebra, only two years older than Semmelweis, had been a pupil of both Rokitansky and Skoda. His career exemplifies the ways in which a bit of original thinking about the possibilities inherent in hitherto neglected medical problems has resulted in the establishing of a new specialty. Hebra is generally regarded as the founder of dermatology.

Since the opening of the Allgemeine Krankenhaus, a group of rooms had been set aside for patients called the *Krätzkranke,* those with scratching diseases. Though these pitiful men and women were visited periodically on rounds, their care was left mainly to the nursing staff. Like certain of the neurological and mental patients, they were pariahs even in the hospital. Because Skoda was a newly appointed attending physician in

1840, these so-called rash rooms were attached for general supervision to his chest department. As has so often happened—then and later—in such situations, Skoda put his youngest assistant in charge of these rooms upon his graduation from the medical school in 1841. The twenty-five-year-old Hebra took his assignment as a challenge.

Like every other disciple of Rokitansky, Hebra was skilled at identifying findings of significance among the complex mass of signs and symptoms exhibited by the confusing aggregation of skin diseases with which he had to deal. Basing his studies on the principles of pathological anatomy, he was so successful at description and categorization—not only in the research but in his teaching as well—that he was given his own department in 1845, only four years after his appointment. It was the world's first dermatology unit.

Hebra was a sensitive, good man, who, in spite of a touch of sarcasm in his much appreciated sense of humor, was well liked by students, faculty, and patients, with whom he seemed to have a particular rapport. He and Semmelweis became close friends. Because of their friendship and his belief in the scientific validity of the Semmelweis doctrine, Hebra would be the first to set forth the puerperal fever theory before an audience of colleagues.

And thus was set the stage for the next phase in the saga of Ignác Semmelweis. His theory embodied the philosophies of the young professors destined to become the leaders of the medical school. It was a standard around which they could rally in their struggle to overthrow the older faculty and perhaps even the government ministers who supported them. Not only did the doctrine's development reflect the investigational techniques that Semmelweis had learned from

Rokitansky, but it followed directly from the reasoning methods he had been taught by Skoda, both of which framed the methodology in the work of Hebra. These three brilliant and visionary young men, along with our hero, were the most significant of the dramatis personae of the Sophoclean tragedy about to enter its most dramatic moments.

VI

Having discovered the nature of puerperal fever, having devised a way to prevent it, and having accomplished these feats of clinical science in a logical, lucid way consistent with the principles of inductive reasoning and the new discipline of pathological anatomy—having done all of this, Semmelweis might have been expected to sally forth into the medical literature with a full description of his contribution. And he might also have been expected to conduct controlled experiments to confirm in the laboratory the theory he had devised in the clinic. He did neither.

The most obvious sort of experiment that suggests itself is a relatively simple one, in which the slightly traumatized genital canals of perhaps a dozen rabbits would be rubbed forcefully with a rod or brush contaminated with infected fluid taken from a woman dead of puerperal fever. Another dozen would undergo the same procedure, except that the contaminated rod or brush would be soaked in a chlorine solution. A third group would also be treated similarly, but with a rod or brush free of contaminated fluid. The proof would consist of the observation that all or almost all of the first group would die, with autopsy findings consistent with puerperal fever; none of the second group would die; few, perhaps none, of the third group would die. But Semmelweis attempted nothing of such a nature.

There was, in fact, one feeble attempt at experimental confirmation, but the results hardly advanced the Semmelweis cause. Unable to hold out against the urgings of Skoda (whose own experience in laboratories was limited), he halfheartedly enlisted the aid of his friend Georg Maria Lautner, a young lecturer at Rokitansky's institute. The two reluctant researchers carried out a series of nine poorly planned experiments between March and August of 1847. In the first seven, the protocol consisted of inserting a brush into the vagina and uterus of an immediate postpartum rabbit, the brush having been soaked in one or another liquid obtained from a cadaver. In the other two, the fluid was injected into the genital canal with a syringe.

In the first three experiments, the material used to saturate the brush was turbid discharge from the abdomens of women dead of puerperal fever. All three animals died, with autopsy findings precisely those found in the cadavers. The other six studies used a variety of fluids taken from several sources, as follows: blood from a man dead of malnutrition, and then chest fluid followed by peritoneal fluid from a man dead of tuberculosis (rabbit remained healthy); peritoneal fluid from the same source as in experiment 4 (rabbit remained healthy); infected chest fluid from a man dead of undescribed cause, followed by peritoneal fluid from a man dead of typhus (rabbit died with findings that were indeterminate but definitely not those of puerperal fever); pus from an abscess between the ribs of a man dead of cholera (rabbit remained healthy); undescribed fluid, injected by syringe into the genital canal of the same rabbit used in experiment 4 (rabbit died with autopsy findings of peritonitis, but not definitely characteristic of puerperal fever); peritoneal fluid from a man dead of undescribed cause (rabbit died with same findings as in experiment 8).

Though Semmelweis alleged that "the changes found in the rabbit cadavers are the same as appear in human bodies as a result of puerperal diseases," this is hardly the case except for the first three animals. As a whole, no definitive conclusions can be drawn from the outcome of these studies. Even allowing for the likelihood that the investigator's bias caused him to overread the result in three of his last four experiments, the most that can be claimed for these nine attempts is that their aggregate outcome is highly suggestive—and a critical twenty-first-century researcher might not concede even that much.

In the last analysis, however, it must be said that this series of experiments, haphazard as they were, did indeed provide strong evidence in support of the *Lehre*. If Semmelweis's overreading of the postmortem findings in rabbits six, eight, and nine is discounted, what is left is the indubitable fact that every one of the three animals exposed to the infected fluid died of puerperal fever. When animals were inoculated with anything else, they either remained healthy or died with a different pathology. It is remarkable that an observer as skilled as Semmelweis should have read more into the misinterpreted autopsies of the three than was actually there. It is even more remarkable that he never made further studies. That this first attempt was poorly designed is understandable. After all, neither of the two experimenters had any training in the methods of laboratory research that were then being developed at the Allgemeine Krankenhaus and elsewhere. But Semmelweis might have turned to one of the several innovators in this area, who would certainly have helped him devise the sort of study that might have provided the definitive proof that was lacking. That he did not is probably traceable to the influence of Rokitansky's teachings. Rokitansky was strictly a descriptive pathologist; he did not do experimental work. Semmelweis very likely had little appreciation of the value of laboratory research and experiments. To him as to his mentor, it was sufficient that repeated demonstration, statistics, and scientific logic led to his conclusions. This and his failure to use the microscope would lead Rokitansky, late in his own career, into serious errors that have somewhat marred his heritage.

Help of a type even more decisive for Semmelweis than a well-planned laboratory protocol would also have been available, through the legacy of Joseph Berres, who had been designated professor of gross macroscopic anatomy at the Vienna Medical School in 1830. Whatever might have originally been meant by this redundant designation, the appointment brought to the faculty a skilled and enthusiastic microscopist who—despite the title of his chair—would make significant contributions not only in histologic research but in the training of several disciples who continued to advance the field. Berres assumed his professorship just as a Vienna optician, Simon Plössl, was developing an improved compound microscope. Coincidentally two years later, a technical difficulty was solved that had been plaguing microscopists for a hundred and fifty years. This was the problem of chromatic aberration, in which a spectrum of colors distorted much of the field being visualized. When an amateur student of lenses, the Englishman Joseph Jackson Lister, discovered the principle that optic physicists call the law of aplanatic foci, he was able to devise a lens system to overcome the distortion. With the combination of the compound microscope and achromatic lenses, the microscope was transformed all at once from being not much more than a scientific toy into being a powerful tool for research.

Semmelweis could have been the beneficiary of the rapidly expanding technology. The thirty-seven-year-old Joseph Hyrtl, a Berres protégé and professor of anatomy since 1844, was one of the outstanding members of the young guard of innovators trying to revamp the medical school and free it from the control of the ministries. Had his obstetric colleague enlisted his help, the two of them might have forged truly remarkable achievements, perhaps even identifying the bacteria that would, years after Semmelweis's death, be proven to be the actual "invisible animal particles" responsible for the deaths of so many hundreds of thousands of women. They might even have anticipated Pasteur.

But this, too, did not happen. Here the reason is once again not difficult to fathom, and is once again the influence of Rokitansky. Accomplished observer though he was, Rokitansky was a naked-eye pathologist, the last in that brilliant series begun with Morgagni. Not only did he fail to do experiments, but he never used the microscope. He in time evolved into a paradoxical figure—continuing to make important contributions in descriptive gross pathology while becoming an anachronism, a pathologist whose work was uninfluenced by microscopy or carefully designed experimental studies. Given that his mentor was uninterested in the wonders to be found in that hidden place which Louis Pasteur later called "the world of the infinitely small," it is no wonder that Semmelweis seems not to have realized how important microscopy was to elucidating the processes of disease, or to verifying his *Lehre.*

And another thing did not happen. Bad enough that Semmelweis did not do proper substantiating experiments; bad enough that he never availed himself of the microscope's help that might have validated his theory in the minds of potential critics; but the worst of his omissions by far was his failure to describe the *Lehre* in the pages of a medical journal.

It was Hebra who stepped into the gap. In the fall of 1847, armed with the results that simple hand washing had achieved, he submitted an editorial to the journal of the Medical Society of Vienna, published in December under the title "Experiences of the Greatest Significance concerning the Etiology of Epidemic Puerperal Fever in Lying-in Hospitals." It summarized the facts leading up to the Semmelweis discovery and concluded with a request for comments and opinions about the experiences at the Allgemeine Krankenhaus, "either to support or to refute them." Though the journal was widely read outside of Vienna, there were few responses.

In April 1848, Hebra published another such editorial, noting that only Gustav Adolf Michaelis of Kiel and Christian Bernard Tilanus of Amsterdam had replied, each with corroborative statements. Once again, he concluded his article with an invitation to contact the journal with comments on the *Lehre.* Michaelis had been informed of it by one of his assistants who, like so many other academic physicians at the time, had traveled to Vienna because it was fast becoming one of the leading centers for medical progress. To test the new theory, the professor made chlorine hand washing compulsory at his hospital, with a gratifying decrease in maternal mortality.

By the time of Hebra's second editorial, almost a year had gone by since Semmelweis achieved his first improved results from the chlorine ablutions. By word of mouth from physicians visiting the Allgemeine Krankenhaus and by way of letters, the message was getting out to the obstetric centers of Europe. Semmelweis himself is said to have written a few of these letters, but there is no documentation of this. Though some of the responses by then beginning slowly to filter back to Vienna were favorable, most were not. Even Tilanus, who had been said by Hebra to support the findings, had certain reservations. He agreed that chlorine prophylaxis was effective and that cadaver poisoning was a factor, but he continued to be convinced that the periodic occurrence of epidemics of puerperal fever meant that atmospheric influences were involved. Accordingly, he did not see any reason to introduce hand washing in his clinic.

The objections to the Semmelweis *Lehre* that were beginning to be expressed at this time foreshadowed resistance it would face in the succeeding years. A contributing factor to this situation was the difference between the meticulous record keeping at the Allgemeine Krankenhaus and the rather desultory methods typical of other institutions. Conditions in Vienna for statistical analysis and clinicopathological correlation were unique, and available nowhere else. In the thirteen years that passed before Semmelweis himself published the details of his observations and conclusions, only those who had watched them develop at first hand were able to gauge their accuracy and follow the detailed logic that had given rise to them and to his ultimate method of preventing disease.

As a result, for example, some facts were confused in the minds of those who read about the doctrine. The problem was manifest in the nature of what was meant by "cadaver poisoning." Though Hebra had called attention to the woman with breast cancer and to the infected knee, these sources of contamination were largely ignored in the emphasis placed on the role of infection transmitted by the dissection of dead bodies. It hardly seemed possible to some critics that there was something unique about a corpse that enabled it to harbor material that could be transmitted to patients. Uncertain about just what was meant, some of the puzzled physicians abandoned the entire theory.

And, of course, there were those who, quite understandably, felt horror at the possibility that they had been killing their patients for years or decades. What must it have been like, to be confronted by a theory that placed the blame for hundreds of deaths squarely on the shoulders of the very obstetricians who were being asked to evaluate its validity in an objective manner? For many, it would prove to be intolerable. Better to convince oneself that it could not be true, as Klein seemed to be doing.

Michaelis, though, knew that Semmelweis was right; to him, there was no way to avoid the truth, though it filled him with guilt. He was to pay a steep price for his enlightenment. Only a few weeks after receiving word of the new findings from Vienna, and before instituting the program of hand washing, he had performed the delivery of a beloved niece, who died of puerperal fever several days later. Overcome with ever-worsening remorse as he watched the mortality drop in his own clinic, he committed suicide in August.

The English proponents of contagionism were, predictably, not impressed by the theory. They had long been saying that something in the atmosphere around a doctor who had been in contact with a case of puerperal fever could be transmitted to his next patient. Practicing simple cleanliness as was done in Britain, they said, was the answer to preventing the disease, not the invoking of mysterious and invisible particles of whose existence no one had experimental or microscopic proof. Semmelweis was frustrated by their inability—or refusal—to distinguish between contagion and infection. Though it seemed to some a tiny and unimportant distinction, it was a major one to him. He would in later years point out that their thesis incorrectly took puerperal fever to be a specific disease accompanied by a specific pattern of pathological abnormalities; it might be airborne or otherwise carried, without the necessity of what Semmelweis called "decomposed organic matter," or what we would nowadays call bacteria-laden pus. In his attempts at refutation, he used smallpox as an example of such a specific disease. Whereas it could be transmitted only from another case of smallpox, the pathological changes of puerperal fever could result from contact with *any* source of pus, be it a sick mother, a putrid cadaver, a lanced boil, or a pus-soaked sheet. Three factors were required for this to happen: the source of putrid material, a means of physically transporting it (airborne means would not do) from that source to make intimate contact with the victim, and an injured surface that allowed absorption, such as the denuded lining of a postpartum uterus or a lacerated finger. As Semmelweis would put it in an 1860 paper published in Hungarian and called "The Difference of Opinion between the English Doctors and Myself about Childbed Fever," "Puerperal fever is a transmissible, but not a contagious disease." Invoking his disease analogy, he pointed out that only smallpox can produce another case of smallpox, and that is what is meant by contagion. An abscessed tooth or an infected uterine cancer cannot cause smallpox. But the pus that exudes from them *can* cause puerperal fever. Neither the English contagionists nor Oliver Wendell Holmes had made that discovery. Semmelweis argued that they did not understand the true nature of the disease or its real method of transmission. He might have added that only he speculated about a specific single cause by which he was identifying a disease, rather than by its symptoms alone. That cause was the decomposed animal matter that in later years would be shown to be bacteria-laden pus. The "invisible particles" were the microbes themselves.

Johann Klein had his own reasons for opposing the Semmelweis doctrine, beyond his constitutional inability to accept anything that did not proceed from his own leadership. From the beginning, he had viewed with alarm the increasing influence of the younger physicians at the medical school, especially because it was being gained without the political clout that kept him in his job. Like everyone else at the Allgemeine Krankenhaus, he was well aware that the *Lehre* was the outgrowth of the teachings of Rokitansky and represented the new approaches to medical theory that threatened his stand against innovation. He was irked by the way Hebra, Skoda, Hyrtl, and others were rallying around Semmelweis, as though his discovery were a powerful armament against authoritarianism, Klein himself, and the conservatism he represented. He was also worried by the rising tide of academic dissent he was witnessing at the university, mirroring the rebellious mood evident among the Austrian populace.

And being human, he was having difficulty facing the increasing evidence that Semmelweis had discovered something truly valuable that might save many lives, something that his own refusal to change an outmoded viewpoint had prevented him from seeing. And if that something was as true as the evidence was every passing month confirming it to be, then Klein himself had been the purveyor of death for thousands of women whose lives were lost because of the methods he had instituted upon taking over from Boër as director of the lying-in hospital. No matter his hard-heartedness to those he considered his opponents, antagonists, and enemies, he was, after all, a physician and deeply affected by the carnage on the obstetric wards. Like many others, he could not face his own culpability. It was easier on the conscience if he did not remove his dark spectacles and take the cotton wool from his ears; this was the time to entrench his position, lest the world tumble about those ears, as Theodor Billroth so well stated it a generation later.

In the midst of all of this, the rebellious mood in the streets of European cities gave rise to a rebellious mob, and before long the uprisings that became known as the revolutions of 1848 were under way. Barricades were erected in main thoroughfares, kings and ministers fled, and all the Continent was clamorous with demands for self-determination and democracy. As always in such outbursts of popular dissent, students and young people in general were in the forefront of the action.

In Vienna, the uprisings reached their height in March,

after Metternich rejected a petition from the diet of Lower Austria asking for political and social reforms. An angered group of students sacked the diet's chamber and clashed with soldiers sent to quell their demonstration. Shortly afterward, a group of students and young faculty formed themselves into a uniformed corps called the Academic Legion, in the hope that a show of unity and perhaps the threat of force might bring about the changes they sought. Students abandoned their books and lectures, and even the compulsively attentive Semmelweis was far less frequently at work than usual. For the first time in the history of the Allgemeine Krankenhaus, there was not a single case of puerperal fever during the entire month, in either of the two divisions. If Semmelweis needed further substantiation of his doctrine, the upheavals of that time provided it. He had long observed that mortality dropped during periods, like holidays, when students did less dissecting than usual, but here was a dramatic example of it. In this case, *post hoc* was truly *propter hoc.*

Semmelweis enrolled in the Academic Legion and proudly wore its uniform (gray trousers, tightly fitted black jacket, and wide-brimmed hat festooned with a large waving plume) while lecturing and often when making his rounds in the First Division. Hebra, also a member of the Academic Legion, later wrote that his friend was wearing the entire outfit when he arrived to deliver Frau Hebra.

As a Hungarian patriot, Semmelweis joined others in personally coming out to meet the revolutionary leader Lajos Kossuth during the third week of March, on his way to Vienna, where he would present his country's demands for autonomy. Activities such as this one, his wearing of the legion's uniform in the hospital, and his statements of solidarity with the revolutionaries did not go unnoticed by Klein, whose store of resentment against his assistant was already considerable and daily growing larger.

As the revolt lessened in intensity and then ended in November, it became clear that there would be no significant political change. But reforms did occur in the academic arena. The Vienna medical faculty achieved much of the autonomy from the ministries that most of its members had wished for. But great benefits did not at first accrue. Though some of the most hidebound of the old guard were forced to retire, Klein, because of strong and continuing connections, was permitted to stay on. Now the controversy between the reactionaries and the reformers came out into the open. The reformers would in time win, but not before their heads had been bloodied by their adversaries.

Meantime, efforts to propagate the *Lehre* continued. In January 1849, Skoda proposed to the faculty that they appoint a commission to study its details, confident that the outcome would be complete acceptance. The great majority favored the proposal, but such was the lingering power of Klein that he and the relatively small group of remaining reactionaries managed to block it. But Skoda persisted. Since he could not persuade Semmelweis to publish, he took all of his young friend's notes and prepared an address to be given to the Vienna Academy of Sciences, the highest scientific organization in Austria. He had to use the notes because Klein had by that time barred all access to his department's records. But Skoda's talk, given on October 18, 1849, emphasized cadaver infection, and once again those who heard it came away with the wrong impression, to be solidified when the remarks were published in the academy's journal late in 1849.

Because Skoda was by then a preeminent figure in European medicine, his writings were taken very seriously by every physician who saw them. Though Semmelweis was elected to membership in the academy following publication, he could not have been pleased that readers and the increasing number of obstetricians commenting on his doctrine misunderstood its basic principles. He was also not pleased that the academy suggested that he perform more laboratory experiments—even offering him a grant to do them in conjunction with Ernst Brücke, professor of physiology. But he refused, as if insulted, stating that there was no point in experiments, since the clinical evidence was so conclusive. Brücke was not put off by Semmelweis's response to the proposal. Wishing to support a thesis of whose validity he was convinced, he contacted Professor Joseph Schmidt of the Charité hospital in Berlin, hoping to make him an advocate. But he made the mistake of writing only of cadaver poisoning. Not surprisingly, the German obstetrician responded that though he agreed that this means of transmission was a way of causing puerperal fever, it was certainly not the only one. This so angered Semmelweis that he would later say of Schmidt in the *Etiology* that the reason he did not support the *Lehre* "did not lie in Schmidt's lack of opportunity for observation, but because Schmidt does not possess the ability to make observations."

Skoda's well-intended address had another unfortunate outcome beyond the misunderstanding it spread concerning the role of cadaver infection. During the course of his discussion, he made reference to the Prague Lying-in Hospital as one particularly beset with the problem of puerperal fever. Pointing out that the dominant theory of its etiology in

Prague was the notion of epidemic influences, he recommended that the chlorine washings be instituted. When the article was published, exception was taken by Friedrich Wilhelm Scanzoni, assistant to the Lying-in Hospital's professor of midwifery, who had only a few years earlier come forth with his own theory, that puerperal fever was caused by changes in the constitution of the blood, caused by cosmotelluric influences. Furthermore, he had stated unequivocally, "By no means is the wound of the uterus represented by the placental site the real cause of origin of the puerperal fever." The thin-skinned Scanzoni took Skoda's comments as a personal attack, and would ever after be an enemy not only of the Vienna professor but of Semmelweis, too. When he assumed the chair at Würzburg a few years later, he was an implacable opponent of the *Lehre*.

About six months earlier, in March 1849, Semmelweis's two-year term as assistant had come to an end. Klein, still clinging to power despite the erosion of so much of his earlier support, refused to renew it. Rokitansky, Skoda, and Hebra urged him to reconsider, but he was adamant. At the same time, they continued to exert pressure on Semmelweis to publish a full report of his work, and this, too, was to no avail.

Semmelweis applied for another term as assistant, as his predecessor had been given (and as his successor would in turn receive), but this was also denied. He appealed to the dean's office, but Klein countered by accusing him of autocratic behavior in the way he kept demanding that students and staff wash in the chloride solution. Anton Rosas, the fifty-eight-year-old professor of ophthalmology and one of the surviving reactionaries, added his voice by claiming that the assistant had to be replaced because the tension between him and Klein was harming the clinic.

Klein carried the day, and Semmelweis found himself without a job. He now had plenty of time on his hands, and his friends pressed him to write, but nothing came of it. He resisted mightily when Skoda importuned him to do the experiments for which the Academy of Sciences had offered financial support, but he would have none of it. He applied to be *Privatdozent* in midwifery, a post equivalent to being a physician in private practice who teaches at the medical school, but the response was slow in coming.

Pushed finally into some action by the continued coaxing of Rokitansky, Skoda, and Hebra, Semmelweis, three full years after identifying the cause of puerperal fever and finding a means of prevention, agreed to discuss his work before a forum of his peers. On May 15, 1850, he took the rostrum at a meeting of the Medical Society of Vienna, to which he had been elected the preceding July through the sponsorship of the trio of supporters, determined to encourage him to continue his efforts. Perhaps by this time he felt safe in speaking publicly because Rokitansky was now the organization's president; in any case, he is said to have handled himself well during the talk and in the ensuing question period. He returned in June for a follow-up address in which he replied to his detractors, primarily Scanzoni. His final lecture took place on July 15, after which he was praised by Johann Chiari, who had been not only Klein's assistant in 1842–44 but was also the professor's son-in-law. Chiari had been one of Semmelweis's teachers when he first began to study obstetrics afer graduating from medical school, and he well knew the younger man's capabilities. Not only that, but he had watched the *Lehre* develop and seen its enormously impressive results. In spite of his relationship with Klein, who was probably seated in the audience that evening, he stood up in support of Semmelweis and his theory. One can only guess what went on at the next family event at the Kleins'.

Semmelweis also received high praise that evening from Dr. Theodor Helm, acting director of the Allgemeine Krankenhaus, who pointed out that the *Lehre* was quite a different thing from the proposals of the English contagionists, and then went on to take issue with the various obstetricians who had stated their opposition to it. Rokitansky concluded the meeting by declaring that Semmelweis had won a resounding victory for his theory. There seemed to be little doubt that he had taken a long step toward its ultimate victory.

At this point, the Semmelweis theory of puerperal fever stood on the verge of acceptance. It had the support of the emerging leaders of Vienna medicine, and its author had shown himself capable of defending it in an open forum and responding with certainty to those who questioned it. Having neglected or refused previous urgings to publish, Semmelweis had been presented with yet another chance to make up for those lost opportunities, and with the ideal timing and influential backing for a well-placed clinical paper. His days were free of professional obligations, and he had plenty of unoccupied hours in which to devote himself entirely to the effort. Though he had lost his job, Semmelweis was nevertheless riding the crest of a wave of appreciation for his abilities and the potential usefulness of his contribution. Just a little more time and effort would have assured him of his rightful place among his colleagues. The proof that this was so, in fact, is to

be found in an engraving of the collegium of the Vienna
Medical School faculty made only three years later, in 1853.
The sixty-five-year-old Klein is inexplicably absent. Of the
fifteen professors pictured, nine had given active support to
the *Lehre* by speaking and writing about it or by overt actions
in its behalf. Only Rosas remained to represent the opposition. Rokitansky's power and influence were such that he had
been elected rector of the University of Vienna the preceding
year. The new guard had triumphed.

But nothing could persuade Semmelweis to write. Because
he did not submit his lectures for publication, they appeared
only as abstracts in the medical society's minutes. Eduard
Lumpe, who had spoken in opposition to the theory, had his
comments published in full. Not only was a great opportunity lost in this way, but those many physicians from elsewhere in Austria and abroad who read the journal were left
with an erroneous idea of the theory and the way it had been
successfully presented and defended at the meeting. This was
not, however, the greatest of the mistakes that Semmelweis
made in the months following that successful evening.

When Semmelweis's appointment as *Privatdozent* did not
come through, he applied again, in February 1850. His application was finally approved at a faculty meeting in March,
and the decision was forwarded to the Ministry of Education.
Along with it went the faculty's recommendation that he be
exempted from the ministry's regular rule that a *Privatdozent*
be restricted in his teaching to the use of "the phantom," the
female anatomical model. The new *Privatdozent* was to be
permitted to teach from the cadaver as well. The ministry
accepted this, but when Semmelweis was notified of the decision in October, the terms had mysteriously been changed

The collegium of the Vienna Medical School faculty, 1853. Of the fifteen professors pictured, nine had given active support to Semmelweis's Lehre by speaking and writing about it or by overt actions on its behalf. Only Rosas remained to represent the opposition. Seated, from left: Franz Schuh, Anton Rosas, Karl von Rokitansky, Joseph Skoda, and Johann Dumreicher. Standing, from left: Joseph Hyrtl, Carl Sigmund, Joseph Redtenbacher, Franz Unger, Carl Haller, Ernst Brücke, Johann von Oppolzer, Theodor Helm, Ferdinand Hebra, and Johann Dlauhy. (Courtesy of the National Library of Medicine)

without anyone's seeming to know who was responsible. His supporters were surprised by the printed schedule of classes for the winter session of 1850–51, with the course announcement reading, "Lectures on midwifery with practical demonstrations on the phantom five times a week by Dozent Ignác Semmelweis." He was to be given no access to cadavers for teaching.

If Rokitansky, Skoda, Hebra, and the others were surprised

by this unanticipated treachery, Semmelweis was stunned. After having waited so long for the approval of his appointment, he had been granted it with a restriction that he took as a personal offense. It was all too much for him: the destruction of his hopes to maintain an unrestricted position on the faculty; the continuing enmity of Klein and his few remaining like-minded colleagues; the realization that his theory—which he was convinced should be not only accepted but hailed, because the evidence was so unchallengeable—was facing continued opposition; the frustration, and affront, too, of not being permitted to teach on cadavers; perhaps a growing concern that his revolutionary activities in 1848 had made him more enemies than he had thought—he would tolerate no more treatment as some sort of academic pariah who needed propping up by more successful associates. Why could everyone not grasp the magnitude of his accomplishment? Why, indeed, was he not recognized as a savior of women? Why, in spite of their efforts to help, did his colleagues continue to badger him with demands for experiments and publication? Why should they require additional proof that he had made a great discovery? How much more could they possibly want?

The cup of bitterness had been gradually filling, and the restrictions of the *Privatdozent* appointment brought it to overflowing. Five days after receiving notification of the ministry's action, he fled Vienna to return to Pest, without so much as letting a single one of his colleagues know of his plans.

His triumvirate of brilliant supporters was shocked. They had given him encouragement and support, and Hebra had become his personal friend. Skoda, who had been more active in promoting the *Lehre* than anyone else and never hesitated to make enemies in its behalf, was enraged by Semmelweis's ingratitude and the manner of his departure. He never spoke the name of Semmelweis again. Even the warmhearted Rokitansky took years to forgive him. Only his generous friend Hebra showed any restraint in his anger. To all of them, it was as though a trusted warrior had deserted in the midst of a decisive campaign that he himself had initiated. Semmelweis was betraying his own cause.

VII

Semmelweis returned to a Buda-Pest stripped of the passion and optimism with which its courageous citizens had embarked on their crusade for autonomy. All the reforms gradually won during the years leading up to the events of 1848 had been wiped away in the failure of the revolutionary upheaval. Political and cultural suppression was the order of the day. The dominance of the empire having been forcefully reasserted, Buda-Pest was governed by puppets as though directly from Vienna, with a harshness that discouraged enterprise or independent thought. Austrian spies had infiltrated everywhere and found plenty of native confederates to join them in their work of ferreting out any activity that might be interpreted as conspiratorial. The regime discouraged free thought and stifled individuality.

The events of 1848 had not improved academic freedom in Hungary, as they had in Vienna. In fact, they served as a warrant to tighten political control at the University of Pest and to add to the repression of its faculty. The government appointed a provisional council to direct the school's affairs, in order to take away what little of the former independence still existed. The council purged the faculty of those thought to be dissidents, discharging several leading professors. A mood of stagnancy overcame not only the university but the small group of well-educated citizenry of the land. At that time, Buda-Pest was still a long way from being the exciting city it later became, from which great literature, philosophy, and science emanated; it was an intellectual backwater. There seemed no motivation or energy to make it otherwise. Politically, economically, and socially, Hungary was a dispirited province of the Habsburg empire.

In no area was the general mood of defeat and pessimism more manifest than in the conditions surrounding the teaching and practice of medicine. The few organizations that had been established during the hopeful days of the late thirties and forties were obliged to close down in 1849, including the Medical Society of Pest-Buda and the Conference of Hungarian Physicians and Natural Scientists. The country's only professional journal, *Orvosi Tàr* (Medical Review), established in 1831, stopped publication shortly after the revolutions. Though the medical society started up again early in 1850, it was an attenuated version of its former self, without sufficient funds or a proper place to meet. Only gradually did it reach the point where it could sponsor lectures or recruit a significant number of members. Not until 1855 was a regular headquarters established and the membership able to rise beyond the ninety to which it had slowly climbed.

The situation was aggravated by the fact that great and rapid medical advances were being made elsewhere at this time, particularly in Vienna. The medical school fell further behind with each passing year, until the level of academic capability was less than mediocre. There seemed no remedy for its backwardness.

Worst of all was the status of obstetrics. It had always been a subordinate department at the school, and students were not required to take its courses or pass an examination in the subject. The obstetrical service had only a small number of beds and delivered no more than three hundred women each year. In its seventeen years of existence, the few obstetric articles published in the now defunct *Orvosi Tàr* had dealt with deformed infants and similar curiosities. The sixty-three-year-old professor of obstetrics, Ede Birly, had written no articles or books. He attributed puerperal fever to impurities arising in the colon, perhaps worsened by the overuse of purgatives. He somehow never made the connection between the school's instituting instruction in pathological anatomy in 1844 and the onset of subsequent outbreaks of puerperal fever so devastating that the lying-in ward had to be periodically closed. Even as the *Lehre* of his countryman Ignác Semmelweis was being talked about in Vienna and other centers, he stuck by his own theory.

In leaving Vienna to return to Pest, Semmelweis had abandoned ship only to be cast up on a desert island. Not only was his professional future uncertain, but his personal life had been markedly changed since his graduation from medical school six years before. His family had been brought to the brink of financial ruin by the events of 1848, and he himself had only small reserves of money. Three of his brothers were in exile or hiding from the authorities, who were pursuing them for their revolutionary activities. Another, Károly, a Catholic priest, had changed his name to Szemerényi in 1844, to display his Hungarian patriotism. And, of course, József and Terézia were dead. The close family circle no longer existed.

Though he was aware of how difficult things would be in Pest, Semmelweis probably had some hope of obtaining a post in which there was an affiliation with the university. But no one came knocking at his door, and he had to be content with the company of a small circle of physician friends who held irregular meetings at which they discussed medical matters. At one of these gatherings soon after his arrival in the city, discussion turned to an outbreak of puerperal fever then prevalent at Pest's 675-bed St. Rochus Hospital. The physicians present knew of the Semmelweis theory, but, like almost everyone else, they erroneously supposed puerperal fever to be caused only by cadaver infection. For this reason, a few of them raised objections to the *Lehre*, pointing out that there was no teaching of midwifery—or anything else, for that matter—at the hospital, and yet the disease kept appearing.

On the following morning, Semmelweis paid a visit to the hospital and soon identified the problem. The obstetric division consisted of a single ward under the management of the chief of surgery. His daily rounds began in his own division and ended in the lying-in unit. In those days, prior to the technological advances that decades later made possible safe abdominal surgery, the great majority of operations were amputations of the extremities or breast, or the drainage of abscesses. The mortality of amputation was close to 50 percent, virtually all of it from infection with organisms so foul that every surgical ward in Europe stank with it. Breast cancers likewise led to infections, because many of them had developed putrid ulcerations before being brought to medical attention. Contemporary descriptions of hospitals tell of the stench that informed a visitor that he was approaching a ward of postoperative patients. But there was another source of contamination as well. In the absence of a separate department of pathology, the regular medical staff was doing the autopsies, providing yet another source of "invisible decomposed organic matter." As if this were not enough, the chief of surgery was also responsible for doing all forensic autopsies, adding to the number of cadaver dissections beyond those of patients who had died in the hospital.

Semmelweis saw his chance to prove himself to his colleagues in Pest while at the same time preventing the death of more mothers. On November 19, 1850, he wrote a letter to the imperial commissioner of health, requesting that he be made unsalaried director of the lying-in ward of the St. Rochus Hospital. Meantime, he had applied for an appointment as *Privatdozent* at the University of Pest under Professor Birly. The university appointment came through during the first week of January 1851, but his request to be made director of obstetrics at St. Rochus took much longer. He was notified of his acceptance in March, but for some unknown procedural reason was not permitted to take over until late in May. He lived on the small remaining patrimony left him by his father, and the funds were rapidly dwindling. He could only hope that starting up a private practice might support him.

The windows of the obstetric ward at St. Rochus looked out over a cemetery, in which Semmelweis knew that more than a few puerperal fever victims were buried. Though the hospital's staff should have had good reason to consider its

rows of graves a constant reproach, they in fact saw no connection between their own activities and the many deaths
being caused literally at their hands. But the new director
knew otherwise. He instituted chlorine-water prophylaxis the
moment he assumed his duties, and was tireless in his personal oversight of every staff member. He had met opposition from students and teachers when he began the hand
washings in Vienna, but his position in the long-standing
hierarchy of authority demanded that his orders be followed.
Things were different at St. Rochus. Here he was an inter-
loper, who had forced his way into the routines of the ward
and insisted that an unfamiliar and seemingly senseless ritual
be carried out before entrance was permitted. More often
than not, he was physically present within yards of the basin
of chloride of lime and never hesitated to harangue a physician or aide who tried to circumvent his procedures. All felt
as though under constant surveillance, and suspicion, too.
And they were right.

With the washings and the absence of the surgeons from
the ward, Semmelweis was able to duplicate the conditions
he had instituted at the Allgemeine Krankenhaus. Mortality
began to drop. Though the prior St. Rochus records—such
as they were—had been lost during the 1848 uprisings, the
puerperal fever mortality had certainly been a great deal
higher than the 0.85 percent achieved under the new rules
during the first four months of the new director's administration. This was precisely the figure attained during the
entire six years that Semmelweis remained at the hospital. Of
the 933 women delivered during that time, 24 died, but only
8 from puerperal fever.

The statistics might have been even better but for an occasional violation of technique, such as the reuse of a contaminated sheet. On one occasion, an assistant surgeon delivered a woman immediately after dissecting the body of a man dead of gangrene, without having stopped at the chlorine basin. Each time such a thing happened, Semmelweis traced it to the source. The successful detective work may have strengthened the reception of his *Lehre* among skeptical physicians, but it hardly endeared him to the implicated members of the staff and those who worked closely with them. His cause was helped not at all by the impolitic harshness with which he confronted the perpetrators.

Semmelweis always viewed his directorship at St. Rochus as temporary, until he could acquire an academic appointment. While building his private practice to a size sufficient to support him comfortably, he was applying to universities within the empire as positions opened up. But it was not until July of 1855 that he was named professor at the University of Pest, upon the sudden death of Birly late the preceding year. Stuck in his ways, the old chief had died unconvinced of the doctrine that his successor would bring with him into an obstetric division where outbreaks of puerperal fever were as common as in hospitals all over the Continent.

In spite of his growing practice and the tiring duties at St. Rochus, Semmelweis had not been intellectually idle during the years since his return to Pest. Within a few months of his arrival, he had joined the newly reactivated Medical Society of Pest-Buda, and the Natural Historical Society, another of the groups that had recently re-formed itself. He became an active participant in their meetings and a well-known member of the city's medical community. He kept in communication with prominent European obstetricians interested in learning more about his work with puerperal fever, and was visited by two of them: Eduard von Siebold, professor of obstetrics at Göttingen, and Bernhard Sigismund Schultze, professor of midwifery at Jena.

Semmelweis was hardly the lonely figure described by some of his more dramatizing biographers, whose depictions of him have long excited the popular imagination to an erroneous notion of him as a misunderstood professional outcast pursuing against all odds the dream of making childbirth finally safe for the coming generations of women.

Of course, he *was* misunderstood, but it was largely his own fault. The primary source of misunderstanding came from the fact that his *Lehre* was based on careful and repeated observations and the meticulous keeping of the sort of records that were available to him at the Allgemeine Krankenhaus. The theory had not arisen solely from the lucubrations of his cogitating mind, nor had it been the result of some inspired insight after years of pondering in a quiet room. Its principles were the result of hard, slogging work at the autopsy table, the bedside, and the library shelf. To fully understand it required a thorough exposition of all the factors that had entered into its development. This Semmelweis had never provided. Though he had been the beneficiary of both opportunity and importuning, he had never sat down with pen and paper to explain the basis of his work. Without giving the kind of detailed analysis he had applied to the problem, he could hardly have expected acceptance on the basis of the rather sketchy general outline then available through the writings of Hebra and Skoda, which focused more on the outcome of the studies than on their evolution. The more sophisticated the mind of a physician considering the theory, in fact, the more evidence would be required to appreciate its validity, which goes a long way toward explaining why some of the leading obstetricians of Europe were hesitant to embrace it. Moreover, the lack of a comprehensive published monograph by Semmelweis had allowed to remain current the widespread misapprehension that he believed the cause of puerperal fever to be infection spread only from the cadaver.

Those biographers and others who for so long thought of Ignác Semmelweis as the castigated possessor of a great discovery that had been suppressed and hidden by resentful colleagues might have changed their opinions had they merely read the description of him presented in a brochure provided to the voting faculty on the day of his election as professor at the University of Pest : "Ignác Semmelweis. Aged 36, *primarius* in Rochus Hospital. In 1846, he became Assistant in the Obstetric Clinic of Vienna. His well-known discovery has received the recognition of the Academy of Sciences of Vienna, and he is considered capable of further research. Member of several medical societies."

That Semmelweis had not entirely destroyed his connections with his colleagues at the Allgemeine Krankenhaus is suggested by a letter that had appeared in the *Wiener medizinische Wochenschrift* (Vienna Medical Weekly) shortly before the Pest professorial election. The Hungarian source is unknown, but the fact that the editors chose to print it would seem to indicate that at least some of his old supporters continued to think highly of him.

It is no indiscretion to mention that both professional and public opinion support the appointment of Dr. Semmelweis to the vacant professorship. . . . Dr. Semmelweis, when

Assistant in the Lying-In Hospital of Vienna, acquired, owing to his lectures and his courses of practical and operative obstetrics, a reputation extending far beyond the boundaries of the Monarchy; and he has already attained a great position in medical practice in our city. . . . If the recently revived scheme of erecting a new lying-in hospital is carried out, and proper facilities for teaching are afforded in it, then will be opened up to our energetic obstetric specialist a wide field of activity and a new era in obstetric science will commence in our fatherland.

It is difficult to know what should be made of the fact that the *Lehre* is not mentioned, in a letter that has the character of a testimonial. Perhaps because the letter's origin was Pest, and also perhaps because its author might have feared stirring up more opposition than help, the *Lehre* was omitted.

The professorial appointment had to be approved by the minister of public education in Vienna and the Hungarian capital's governing body, the Presidential Council of Buda. Such was the political climate of Hungary that the council referred it to József Prottmann, the Pest chief of police. Prottmann reported that the elected candidate was a Hungarian patriot who had not engaged in political activities since 1848; he was in every way a suitable person for the post. On July 18, 1855, Semmelweis was officially appointed by the emperor to be Professor of Theoretical and Practical Obstetrics at the University of Pest.

At first, Semmelweis tried to hold on to his directorship at St. Rochus, hoping to do some gynecological studies there in order to complement his obstetrical work at the university. But one of the unsuccessful candidates for the Pest professorship, Ignác Rott, petitioned the government against allowing a man to hold two positions, one responsible to the municipality and one to the state. Accordingly, Semmelweis had to resign effective June 1857. His successor was the selfsame Dr. Rott.

Another reason for Semmelweis's attempt to continue in his post at St. Rochus was the inadequacy of the lying-in facilities at the university. They were a far cry from those to which he had been accustomed at the Allgemeine Krankenhaus, and were in fact inadequate for teaching. Though the school's entire medical facility was crowded into much too small a group of clinics, the situation in obstetrics was particularly bad. The entire unit held twenty-nine beds, of which only three were in the ward reserved for women in hard labor. This room had a single window, and it, too, looked out on a burial ground. There being no obstetric lecture room, the new professor was obliged to teach in the corridors or in any available room he could find on any given day. In that first summer of 1855, he was unable to reorganize the course and found himself delivering a series of lectures to a combined class of twenty-seven medical students and ninety-three midwives, all of them squeezed together in whatever space he could find wherever he could find it. In spite of rumors—and even some promises—of a new lying-in hospital, none was ever built.

Semmelweis's situation was worsened by the attitude of the staff, both professional and lay. He found himself resented even more than he had been at St. Rochus, where his arrival had brought with it at least the small advantage of relieving the ward personnel from the control of surgeons. His impatient and verbally abusive response to laxity alienated many who might otherwise have been disposed to ally themselves with a more tactful chief. It soon became obvious that some of the physicians and nurses were doing their best to circumvent his rigid rules regarding cleanliness and chlorine, and he needed to be even more watchful here than in his preceding directorship. He also had to keep an eye on supplies and expenses. The abstemious administration had ignored Professor Birly's repeated requests for more linens, and Semmelweis inherited a problem about which he quickly decided it was useless to continue complaining. Frustrated by the impossibility of getting any response from the hospital's bureaucracy, he bought a supply of linens himself. Predictably, he had great difficulty obtaining reimbursement.

He fired off memorandum after memorandum to the administration, without avail. There was not enough equipment, instruments were too few, and the laundry kept complaining that the obstetric unit changed bed linens more frequently than the regulations called for. But Semmelweis would never give an inch. Regardless of how many of the staff were alienated, he insisted on strict compliance with his orders. Though the results bore him out, he seems to have been able to recruit few allies. In the first year, only 2 women died of puerperal fever among 514 delivered, an unprecedented mortality of 0.39 percent.

Before long, noises were heard from Vienna. When the excellent puerperal fever statistics of the academic year 1855–56 were finalized, József Fleischer, one of Semmelweis's assistants, sent a brief report to the *Wiener medizinische Wochenschrift*, describing the results that had been achieved. It was printed, but with an editorial note appended, reading, "We thought that this theory of chlorine disinfection had died out long ago; the experience and the statistical evidence of most of the lying-in institutions protest against the opinions expressed in this article: it would be well that our readers should not allow themselves to be misled by this theory at the present time."

There is no way to know who wrote this snide editorial that dashed whatever hopes Semmelweis still nurtured about the status of his doctrine in Vienna. If he had thought that his supporters would continue to publish articles promoting it, he was as misguided as they had been in expecting that he himself would publish during those active days at the Allgemeine Krankenhaus. He should have known that the abruptness of his departure had so angered the associates to whom he was most indebted that they were unlikely to continue to aid his cause. And by then he also knew that one of his most outspoken opponents—his main rival, in fact—Carl Braun, had succeeded Klein as professor on the old reactionary's recent death.

Braun had followed Semmelweis as assistant in Vienna. His worldview and Klein's proved so compatible that he had been allowed to remain in that capacity for five years, though his predecessor had been curtly dismissed after only two. When Birly died in 1854, Braun applied to be professor at Pest, and was indeed the faculty's first choice of the six candidates, with Semmelweis being second. But the Presidential Council of Buda, refusing to appoint anyone who could not speak Hungarian, chose the native son. In any event, Braun would not have stayed long in Pest, because Klein's unexpected death opened an opportunity for which he was the obvious candidate. Semmelweis did apply, but it was an exercise in futility, even though the puerperal fever rate had once again risen when his rival took over the wards as assistant at the Allgemeine Krankenhaus.

Within less than a year, Semmelweis was becoming discouraged. He had taken up his duties at the university with the enthusiasm of a man who has at last found the proper stage from which to convince the medical world of the correctness of his theory. But everything was militating against his success. The dreadful and unchanging conditions in the hospital, the antagonism of the staff, the refusal of the administration to provide appropriate supplies and equipment, the glaring loss of support from Vienna—nothing was going right. Nothing, that is, except that he was indeed saving lives. Here, as at the Allgemeine Krankenhaus and St. Rochus, the mortality statistics bore him out in every way. And yet, few became convinced of his doctrine. He was encountering reports in which some of the leading obstetricians of Europe were questioning his claims and arguing against conclusions that he believed to be so evident that to ignore them was to condemn countless women to death. As these frustrations mounted, he grew more determined to justify himself.

Semmelweis had never cared a whit about his personal image in the eyes of others, especially those he considered fools not to be borne gladly. Holding first rank among the corps of fools were hospital administrators. The clashes between idealistic physicians and those responsible for the orderly running of a hospital are legion and the stuff of many a locker-room story, but the lengths to which Semmelweis went in order to make his point are surely unique. One such episode followed hard on the heels of the successful year of 1855–56.

Shortly after the publication by Fleischer, Semmelweis had

a few unexpected deaths in his department. He began to watch over every detail even more intensely than before, as the toll slowly mounted until he had lost sixteen patients during the academic year 1856–57. What made these mortalities different from any he had observed in the past was that not a single newborn became sick, of all the mothers who died. This would make sense only if the infections were occurring after delivery, when the mother was recovering. But how could this be, if no healthy mother was examined by students or physicians once her child had been born? Semmelweis's previous experiences with bed linens provided the answer. To save money, the hospital had contracted the washing to a slipshod laundry at a low price. The head nurse on the obstetric division was accepting laundered sheets still so dirty that they reeked of the decomposed discharges of prior patients. When nothing changed after Semmelweis forcefully complained to the authorities, he determined to take desperate action. Gathering up as large a bundle of the dirty sheets as his arms could carry, he stomped off to the office of the administrator in charge of such things, one von Tandler, and threw the stinking lot on his desk. Though he achieved his objective of having the laundry contractor changed, he had embarrassed von Tandler and his minions, an insult not soon forgotten. From then on, the offended administrator became an implacable foe, encouraging junior staff members to bring him news of any possible irregularities on the obstetric division. Moreover, the head nurse had been made to look foolish, thus yet another enemy had been created.

The grumbling about the professor's unorthodox and highly irregular ways of doing things was only increased by such episodes. As obstinate as certain members of the staff insisted on being, that was how obstinate Semmelweis was in the demands he made on them; as lax and uncaring as others were, that was how compulsive he was in forcing exactitude on them. Always intolerant of disagreement, he indulged himself in outbursts of anger that grew more explosive with each incident. He gave increasingly free rein to a rapierlike sarcasm with which he skewered opponents and obstructionists. Semmelweis became more and more unpopular. And he did not care.

If Semmelweis thought he had solved the problem of dirty sheets, he was wrong. The following year, the puerperal fever mortality was even higher, reaching 4 percent. Though lower than the figures for most of the large clinics of other European countries, it was nevertheless greater than the Hungarian authorities had come to expect. Semmelweis found himself in the unique situation of being officially reprimanded by von Tandler and others for the unacceptable results that were occurring despite the purchase of adequate linens and the improved laundering. The written statement noted, "Confidential communications have been received concerning several unfortunate occurrences and abuses at the Obstetric Clinic of the Imperial Royal University," and went on to castigate Semmelweis for not solving the problem of soiled linens even though the institution had provided him with a considerable amount of money to buy new ones. It soon became clear to Semmelweis that members of the nursing staff had continued to be lax in the way they changed bedclothes. It is impossible to believe that it did not occur to him that his efforts were being sabotaged, and very likely they were. The figures improved as soon as the head midwife was dismissed. In the 1860–61 academic year, not one woman contracted puerperal fever. Meanwhile, in October and November of 1860, there were ninety-six deaths of the disease at the Allgemeine Krankenhaus, on the obstetric service of Semmelweis's critic, Carl Braun.

In spite of detractors, some of whom quite obviously did what they could to prevent his success, Semmelweis had a small group of loyal friends on the faculty. Chief among them was the internist Lajos Markusovszky, with whom he had been particularly close since medical school days and who understood him as did no other colleague. He had also developed a warm relationship with János Balassa, the professor of surgery, and with the microscopic anatomist Ignác Hirschler, and there were others. These included leaders of the medical society, who in 1857 founded a new journal published in Hungarian, *Orvosi Hetilap* (Medical Weekly). To them Semmelweis was a welcome gadfly, but to most of the faculty at the medical school he was a troublemaker who insisted on having his way. For such mediocre men, the status quo was comfortable, and most educated Hungarians had learned, especially in 1848, that attempts to change it were counterproductive and even dangerous. To them, a university post was a sinecure, not to be tampered with.

And in the matter of "most educated Hungarians": most educated Hungarians spoke fluent, unaccented Hungarian. This was not true of the professor of obstetrics at the University of Pest. But he was a patriot and committed to speaking the native language even if it was not his own. As was noted in an earlier chapter, he had learned it in secondary school, and not very well. He spoke the tongue of his own country with a decided German accent, which hardly endeared him to the nurses, the nonprofessional staff, or most of his faculty colleagues. German was the language of the oppressor, Austria. No matter that he frequently dressed in Hungarian national costume for his lectures; there must surely have been many students and staff who wondered where his allegiances lay.

Semmelweis was helped not at all by his personality, which was the opposite of jovial or even friendly. He was a man of determined intensity, and his speech contained few words that had any function save the expository. It was devoid of small talk, superfluous pleasantries, and flowery adjectives. He had long lost that "playful, jocular nature" remarked upon by those who knew him before he set out on his puerperal journey. Neither "lighthearted" nor "popular" any longer, he had developed a seriousness of purpose that excluded trivialities and even the mundane. Other than to his small circle of friends, Semmelweis was not a social—or even sociable, except when it suited him—being. Worst of all, his hackles were raised by every challenge to his increasingly self-righteous certainty.

And then, in 1856 at the age of thirty-eight, the bald, stocky, single-minded obstetrician surprised his small circle of friends by falling in love with a young woman nearly half his age, the twenty-year-old Maria Wiedenhoffer, the daughter of a prosperous Buda-Swabian merchant. Father Károly Szemerényi officiated at his brother's wedding in June of the following year. Between 1858 and 1864, five children would be born of this marriage, the first two of whom died in infancy. (The only surviving son, Béla, committed suicide in 1885, when he was twenty-two years old.)

The marriage brought some contentment into Semmelweis's turbulent life. His increasingly frantic efforts to make others see what he saw were refusing to bear fruit. He was involving himself in a whirlwind of activities—multiple faculty committees, multiple projects, multiple conflicts with colleagues. He was impetuous, he lacked tact, and he had a knack for alienating important people. The serenity achieved at home in the evenings disappeared early each morning when he went off to the wards.

One can only imagine what it must have been like for Semmelweis to read the medical journals of the time, and occasionally to find an article attacking his *Lehre*. Always, the author wrote of the incorrectness or inadequacy of a theory that attributed puerperal fever only to the transmission of infection from the cadaver. And always, the elements of the theory had been derived at second or third hand, from word of mouth or a review of the brief descriptions written years earlier by Hebra and Skoda. Semmelweis dispatched a series of letters to some of the leading obstetricians of Europe, less to correct misperceptions than to solicit their opinions of his work, perhaps seeking to reassure himself in the face of the unsatisfactory articles he was reading. He was rarely pleased by the responses.

By this time, it appears, Semmelweis had reached a point where his sense of self had become so interwoven with his theory that its acceptance or rejection was equivalent to his own. The process had no doubt begun in Vienna, but it was now rising to a climax. The frustration of his inability to convince the medical world of a principle that seemed so obvious and had actually been proven in practice was becoming increasingly difficult to bear; the storm he aroused almost everywhere by his brusqueness and tactless methods had deprived him of the possibility of enlarging his small group of allies; his certainty only made him more self-righteous and provided the justification for the obnoxious forcefulness that alienated so many of his colleagues. Attacks on his doctrine had become attacks on himself. The burden grew greater with each passing month. He felt beleaguered.

Confronted with the virtually universal misunderstanding of his work, Semmelweis realized that the time had finally come when he had to throw off his antipathy to writing. It was imperative that he explain his thesis in great detail and respond to every argument of those who had written against it.

But he was not yet ready to try writing a complete work in the German language. He began his campaign by testing the waters in Hungarian. In his first public lectures on the *Lehre* since those of 1850 in Vienna, he addressed the Medical Society of Pest-Buda on January 2, 1858, and then again on January 23, May 16, and July 15. Markusovszky, present at each talk, was immensely pleased to see that his gruff friend could be as passionate at the rostrum as he was in every spontaneous conversation in which puerperal fever had been discussed since his return to Pest in 1850. He would later write, "Semmelweis expounded his teaching with such conviction before our society that only a man is capable of possessing who not only can fight for its truth but vouches for it with his life. His dedication to his work was evident at the meeting of the Medical Society and it deeply moved all those present."

The lectures were published that same year, as seven installments in *Orvosi Hetilap*. Perhaps seeing his work so well expounded in print gave their author the courage he needed to begin work on a full-scale book with these articles as their framework. Since Hungarian is to this day a tongue rarely spoken by anyone other than Hungarians, the volume would have to be in German, which had in the preceding decade or more displaced French as the language of communication in the medical world. Reluctant as he had been, and insecure in his ability to write with the polish and style of his former Viennese colleagues, Semmelweis in the spring of 1859 finally undertook the necessary chore of sitting down to compose. A year and a half later, in October 1860, the book—*Die Aetiologie, der Begriff und die Prophylaxis des Kindbettfiebers* (The Etiology, the Concept, and the Prophylaxis of Childbed Fever)—was complete, though it would not be officially published until 1861. On its pages were spread for the world to see and wonder at not only the Semmelweis *Lehre* and its justification but the life and soul of its author.

VIII

*T*he *Etiology, the Concept, and the Prophylaxis of Childbed Fever* is a complex book, written in a complex manner by a complex man. One could not be criticized, in fact, for replacing "complex" with "odd." The complexities and oddities of Ignác Semmelweis are reflected in his magnum opus and the way he went about writing it.

For approximately eighteen months, Semmelweis worked furiously at his project, dashing off page after page in the greatest haste, sometimes—especially when responding to the critical writings of those who denigrated his work—inserting paragraphs and entire sections inappropriately because he was at a barely related place in the writing when something he cared urgently to address came to his mind. Even more damaging to sustaining interest was his compulsive need to add new chapters even when he had already covered the same ground earlier. He would insert these without reference to the previous material, decreasing even further the reader-friendliness of the text. After the years of delay, he was so eager to publish his book that he seems never to have reviewed the text as it was taking shape. Certainly, there was no attempt to edit, either by the author or anyone else. Only the printer stood between pen and audience. The result is 543 pages of a book that is logorrheic, repetitious, hectoring, accusatory, self-glorifying, sometimes confused, tedious, detailed to the point of aridity—in sum, virtually unreadable. The difficulty of following its often convoluted sentences in the original German is magnified when a literal translation into English is undertaken.

The *Etiology* consists of two parts. The first is a recounting of the *Lehre* and a chronicle of its development, in which no possible factor, circumstance, datum, or pertinent element of historical background is omitted. The second is a passionate defense against the reservations and criticisms raised by opponents, and an appreciation for the few who agreed. The total of such obstetricians and others is twenty-eight, cataloged and dispatched one by one in a section of 208 pages, the last 6 of which are aimed at that old nemesis, Carl Braun. As a subscriber to the dictum that the best defense is a good offense, Semmelweis goes much farther than merely justifying his doctrines; he never hesitates to level charges against some of the most prominent obstetricians of the time, calling them ignoramuses for not subscribing to his teachings. And worse yet—they are accused of being unrepentant murderers of the women whose lives have been entrusted to them.

Semmelweis wastes no ink on flowery preamble. On reaching the final paragraph of his brief preface, the reader is already aware of having joined an unusual man on a personal journey toward validating not only his theory but himself. The self-exalting author clearly—and worryingly—sees himself as a martyr to a great cause, in whose name he adopts an almost Christlike attitude toward the mission of his life: to bring the Puerperal Truth to a world of unbelievers. And he does it in his inimitable tortured and convoluted syntax:

> Faith has chosen me as an advocate of the truths which are laid down in this work. It is my imperative duty to answer for them. I have abandoned the hope that the importance and the truth of the facts would make all conflict unnecessary. My inclinations are of no moment alongside the life of those who take no part in the dispute over the justice of my claims or of those of my adversaries. I am constrained to come before the public once more, since my silence has been futile, and despite the many bitter hours which I have suffered, yet I find solace in the consciousness of having proposed only conclusions based upon my own convictions.

Having finally published his book, Semmelweis lost no time in sending copies to leading obstetricians and medical societies all over Europe. Only rarely was he satisfied with the response he got. Some of the *Etiology*'s recipients ignored it, no doubt believing that they would encounter nothing new, in view of their presumed familiarity with the misleading decade-old writings of Hebra and Skoda. Others simply did not care, remaining confident that they already had a tenable theory about the origins of puerperal fever, and no contributions by this Hungarian agitator would change their fatalistic attitude about the inevitability of recurrent epidemics. Still others simply could not face the daunting task of making their way through the convoluted prose and the repetitiveness of the frequently bombastic text; they might read a few pages and then give up. And beyond these were the readers who were so offended by the fury of insults directed toward men for whom they had great respect that they rejected the book as little more than an angry polemic. Some in this last group, including several of those attacked, took up the pen to strike back at Semmelweis and his doctrine.

Semmelweis had assumed that the publication of his book would mark a turning point in the acceptance of the *Lehre*. So deeply absorbed was he in the scientific validity and even the moral righteousness of his teachings that he was oblivious to the daunting problems facing anyone making a serious attempt to follow the arguments he was presenting. He further presumed that those whom he was so vigorously attacking would be exposed in his pages as the dangerous promulgators of error that he believed them to be. Not only did none of this come about, but the publication actually worsened his position; it presented his opponents a highly visible target against which to launch their assaults. Semmelweis could no longer console himself with the belief that his work was being rejected only because it had been misunderstood. Here it was, laid out in every detail for all the world to see, and the great majority of its intended audience had turned its collective back on it. Though some major figures did respond with support, all but a very few of these were men who had already expressed their confidence. Not only was there no large-scale conversion to the notion of invisible decomposed animal matter, but the *Etiology* served only to stiffen resistance to it.

Semmelweis was exasperated and enraged. In a monomaniacal frenzy of desperation and dashed hopes, he now lashed out as though against a world dominated by his professional enemies. In a series of open letters, he took on three of the most prominent and adamant of them and then, in a final burst of outrage, declaimed against the entire galaxy of obstetric luminaries who were treating his theory and himself with such disdain. With the same heedless haste that had contributed so much to the failure of the *Etiology,* he fired off his missives both personally and publicly, perhaps having convinced himself that an explosion of wrath would turn the battle his way. Or it may have been something quite the opposite that made him take such a self-defeating step: recognizing the impossibility of victory, he could at least damage his foes. In either case, the air of recklessness detectable in certain sections of the *Etiology* here became the underlying theme of virtually every paragraph he wrote.

To Joseph Späth, professor of obstetrics at the University of Vienna's Joseph's Academy:

Herr Professor, you have convinced me that the Puerperal Sun which arose in Vienna in the year 1847 has not enlightened your mind, even though it shone so near to you. . . . This arrogant ignoring of my doctrine, this arrogant boasting about [your] errors, demands that I make the following declaration:

Within myself, I bear the knowledge that since the year 1847 thousands and thousands of puerperal women and infants who have died would not have died had I not kept silent, instead of providing the necessary correction to every error which has been spread about puerperal fever. . . .

And you, Herr Professor, have been a partner in this massacre. The murder must cease, and in order that the murder ceases, I will keep watch, and anyone who dares to propagate dangerous errors about childbed fever will find in me an eager adversary.

In order to put an end to these murders, I have no resort but to mercilessly expose my adversaries, and no one whose heart is in the right place will criticize me for seizing this expedient.

To Friedrich Scanzoni, professor of obstetrics at Würzburg, who, with the possible exception of Carl Braun, was the most influential of Semmelweis's opponents:

Herr Hofrath [literally meaning "Court Councilor," but used to designate high-ranking officials, such as distinguished professors] will have deduced from my letter to Professor Späth that in order to make an end to the murders I have seized the unshakable resolve to relentlessly oppose everyone who spreads errors about puerperal fever.

Your teachings stamp the physician as a Turk, who in fatalistic, passive resignation allows this tragedy to engulf his puerperal women. . . . You see, Herr Hofrath, I have deprived your teachings of the basis which you found in the murderous deeds which were committed as a result of your ignorance by the midwives and physicians of Würzburg and its vicinity. . . . In this regard, you have, Herr Hofrath, sent out a significant contingent of unwitting murderers into Germany. . . . I have, in fact, devoted 103 pages (pages 315–417) of my publication on childbed fever solely to the refutation of all the errors and deceptions which hold you in their spell. . . . Should you, however, Herr Hofrath, without having disproved my doctrine, persist in writing and permitting to be written about epidemic childbed fever—should you, however, Herr Hofrath, without having disproved my doctrine continue to train your pupils in the doctrine of epidemic childbed fever, I declare before God and the world that you are a murderer and the "History of Childbed Fever" would not be unjust to you if it memorialized you as a medical Nero, in payment for having been the first to set himself against my life-saving theory.

In a second letter to Scanzoni he added,

> To you, Herr Hofrath, there remains nothing else left but to adopt my doctrine, if you still want to salvage something of your reputation, whatever of it is still left to salvage. If you continue to adhere to the doctrine of epidemic childbed fever, advancing enlightenment will cause pseudo-epidemic childbed fever and your reputation to disappear from the face of the earth.

Semmelweis wrote in a similar vein to Eduard von Siebold, professor of obstetrics at Göttingen. And then in one final, all-encompassing display of epistolary fireworks, he composed his final open letter in 1862, addressed "To All Professors of Obstetrics," saying much the same things as he had to the three individuals during the preceding year. He was now taking on everybody.

During the period of the writing of the *Etiology*, it had become obvious to those around Semmelweis that his physical health was failing. The illustrations on pages 164–65 compare his appearance at the age of thirty-nine, in 1857, with that in the year he published the *Etiology*, 1861. At the same time, he had begun to exhibit the worrisome symptom of alternating periods of depression and elation, following no specific pattern. He would bring up the subject of his *Lehre* at every opportunity and verbally attacked anyone who disagreed with him. His closest friend, the internist Lajos Markusovszky, urged him to retreat from the puerperal fever wars and devote more time to gynecology, which had always interested him. He seems to have been surprisingly willing to do this, though he refused to stop vigorously defending his position and seeking out every opportunity to force it into conversation.

By the end of 1862, it was impossible for Semmelweis's colleagues and friends to ignore his increasingly erratic behavior. He was moody, irritable, grandiose, forgetful, and argumentative, sometimes all at once. Hyperactivity and lethargy came and went without obvious cause. He was often sleepless, getting up and roaming the house or even the streets as though determined to carry out some specific task, and talking to himself or imaginary others while about it. At times he seemed quite childlike, and at others he was blustery and irrationally demanding. Angry and insulting one moment, he would in the next embrace and even kiss the object of his outburst. He seemed uncaring about money and inattentive to spending habits and expenses. His previous meticulousness of dress gave way at first to carelessness and on occasion to slovenliness.

Still, manifestations like these were not recognized as pathological, as so often happens when the changes are gradual and the subject is known to be under a great deal of stress. But by the early summer of 1865, Semmelweis would more frequently be heard to make inappropriate remarks, some of his conversations even spiced with gratuitous sexual themes or obscenities. He seemed consumed by sexual desire, sometimes masturbating as soon as he had had intercourse with Maria. Not only that, but the previously rigidly moral and devoted husband began consorting openly with a prostitute as though it were the most normal thing in the world. Somehow, he managed to fulfill—albeit barely—the responsibilities of his professorship, but the gradual deterioration was so marked that those observing him recognized that the situation could not long be allowed to continue.

Matters came to a head on July 21 of that year. József Fleischer, the assistant who had written the letter to the *Wiener medizinische Wochenschrift* in 1856, described the last faculty meeting attended by his professor. During a discussion concerning the filling of a vacant lecturer's post in his department, Semmelweis was called upon to deliver a report about the progress being made. He shuffled to his feet as though distracted, pulled a bit of paper from his trousers pocket, and proceeded to read the full text of the midwives' oath, as though he had no idea where he was. His colleagues could delay no longer. They took him home and saw to it that he was put to bed by Maria, who now had to face the reality she had for so long been denying despite the abundance of evidence: her husband had lost his sanity.

With the help and advice of several professors at the medical school, Maria Semmelweis tried to care for Ignác at home, but her efforts availed nothing. When his condition deteriorated even further during the next week, the doctors convinced him that he needed to recover his health at the spa in Grafenburg. But it was only a ruse, Frau Semmelweis having consented, at their urging, that he be taken to a mental hospital. Accompanied by his assistant, István Bathory, and

Ignác Semmelweis, 1857. (Courtesy of the National Library of Medicine)

her uncle, she set off with him on the evening of Saturday, July 29, on the overnight train to Vienna, from where, he was told, they would proceed to the spa. Little Antonia Semmelweis, because she was still unweaned, was brought with them, five days after her first birthday.

Ignác Semmelweis, 1861. (Courtesy of the National Library of Medicine)

When the party arrived at the Vienna station early next morning, the kindly Hebra was standing on the platform, waiting for his old and lost friend. He suggested to Semmelweis that he briefly interrupt his journey, because he wanted to show him the private sanitarium he had recently opened. Instead, he and Maria's uncle took him to a large state-run insane asylum, as such institutions were in those days called. While Semmelweis stood in the hospital's garden, talking somewhat frenetically with one of the staff members, the two men unobtrusively left. The commitment papers had been signed by the doctors in Buda-Pest, but the reason for admission to a public facility is unknown. Very likely, though, Semmelweis's recent profligacy and inattentiveness to expenses had so eroded his financial reserves that private care was unaffordable.

Forty years later, Maria Semmelweis described her attempt, in the company of her uncle, to visit him next day: "Hofrath Riedel, director of the sanitorium, met us in person to say that the night before my husband had tried to get out and, when he was forcibly restrained, fell into a fit of delirium so that six attendants could scarcely hold him back. I was not allowed to see him."

Two weeks later, on August 14, his family was notified that Semmelweis had died the preceding day. His body was transferred to the Pathological Institute at the Allgemeine Krankenhaus, and there it was autopsied on one of the tables where he himself had so often worked, and where Kolletschka's body had been opened almost twenty years before. The procedure was performed by one of Rokitansky's assistants, Gustav Scheuthauer. Though Scheuthauer signed the report, the neurological studies were done by Theodor Meynert, a Rokitansky disciple who dissected all brains at the asylum. Frau Semmelweis was told that a wound of the middle finger of the right hand, said to have been sustained during a recent gynecological operation, had become septic, with the infection spreading through the bloodstream and causing, among other manifestations, a large collection of pus to accumulate in the chest. In a final touch of grim irony, the autopsy demonstrated findings very like those which had helped Semmelweis solve the riddle of the etiology, the concept, and the prophylaxis of childbed fever. The greatest of the Semmelweis biographers, the University of Manchester obstetrician William J. Sinclair, would be moved to write, in 1909, "[H]e died from that disease to the prevention of which his whole professional life had been devoted—the disease which had carried off his friend Kolletschka, and put himself on the track of his discovery." Thus was added an almost poetic symmetry to the legend of the genius of puerperal fever.

The details of Semmelweis's incarceration in the asylum are unknown, though the Hungarian Society for the History of Medicine possesses photocopies of documents relating to his final days and death, presented to them by a physician, Georg Silló-Seidl, who obtained the papers from the Viennese archives in 1977. The summary of his hospital course is so riddled with inconsistencies, obvious errors, and suspect alterations that it must be considered unreliable in the extreme, very likely having been compiled after its subject's death, which is Silló-Seidl's conjecture. One might add that its inconsistency with the later statement of Frau Semmelweis adds to the suspicion that it was meant to hide certain events that actually took place.

The evidence for what these certain events were comes from the official autopsy protocol that Silló-Seidl obtained from the Viennese authorities along with the other documents; it is supported by the examination, photographs, and x-rays taken of the remains at their disinterment and reburial in Budapest in 1963, and also by Maria Semmelweis's account of her visit on the day following her husband's commitment. When I first saw these various pieces of evidence in 1977–78, the actual cause of death seemed so obvious that I could only wonder why it had not been suggested earlier. It was obvious, in fact, even in Scheuthauer's original report of 1865, or rather his report as the Viennese authorities made it public at that time. The appearance of the body makes it virtually certain that Frau Semmelweis's description of events is accurate. Like so many other wildly psychotic patients of the time, her husband was severely beaten by asylum personnel trying to restrain him shortly after his admission. Injuries to the left hand, four fingers of the right hand, both arms, and the chest are so suggestive that no other conclusion is tenable. The injury to the left chest in particular leaves an observer with the conclusion that the victim of such a brutal assault was stomped as he lay on the ground. It consists of an abscess, visible on first inspecting the corpse as "discolored green skin," under which bulged "a half-sphere swelling like an air-pillow." On cutting into the body, the large bulging protuberance was found to be caused by an extensive collection of "yellow-green pus . . . mixed with stinking gases" located between the chest muscles and the rib cage, encompassing the entire area from the first to the sixth ribs in the front, with an opening where the pressure of the abscess had caused it to perforate into the thoracic cavity, producing a connecting abscess "the size of a man's fist," and reaching as far as the pericardium, the envelope around the heart.

Despite such powerful evidence, I sought consultation with two pathologists, telling neither of them anything of the history or of my diagnosis. Independently, both concurred that such findings could result only from the trauma of an assault.

When I first presented these conclusions in a lecture in February 1978, and then in an article published in the *Journal of the History of Medicine and Allied Sciences* in 1979, they were greeted with skepticism by the few students of Semmelweis's life from whom I later heard. But that is no longer the case. One of America's most authoritative Semmelweis scholars, K. Codell Carter of Brigham Young University, has repeated the judgment in a 1995 article in the *Bulletin of the History of Medicine*, quoting his Hungarian colleague István Benedek as his source.

Of course, none of this explains Semmelweis's mental deterioration. Though the official cause of death was listed as "paralysis of the brain," that was a term meaningless even at the time. The diagnostic summary of the autopsy states that "cerebral atrophy" and "chronic hydrocephalus" were found, but these are observations made with the naked eye, and there is no way to know precisely what is meant by such designations. Moreover, the body of the report is worded somewhat differently, in that the atrophy is said to be in the anterior lobes and hydrocephalus is not mentioned, merely that there is "an ounce of clear serum" in each of the cavities of the brain. No microscopic studies were done on any of the tissues. The result is that the favorite indoor sport of Semmelweis scholars and other enthusiasts for the better part of a century and a half has been the diagnosis of his organic brain syndrome. The great majority of them have seemed content with calling it syphilis.

When I first began studying the life of this tortured medical pioneer, in 1977, I found myself having difficulty reconciling much of his bizarre behavior with what I knew of tertiary syphilis. Seeking help, I consulted Elias Manuelides, then director of neuropathology at the Yale School of Medicine. Telling him only the patient's age and that he had died in the mid-nineteenth century, I recounted the symptoms characterizing the final years of his life. Dr. Manuelides felt sure that the diagnosis of syphilis was not justified by the available information, and suggested a far more likely disease, one that was not described until 1907. He called it Alzheimer's presenile dementia; it was completely consistent with everything I had learned of my subject. Particularly impressive was a clue that I had not yet discovered, namely the marked aging exhibited by the patient over the course of only four years, a distinctive characteristic of this condition.

Some explanation is in order here. Dr. Manuelides made this diagnosis a year or more before Alzheimer's—until then mistakenly believed to be a rara avis occurring only in youngish or middle-aged people—was shown not only to be a disease most commonly found in the elderly but also to be far more prevalent than had been thought. As a medical student in the 1950s, I learned about it in four sentences of my 1,400-page pathology textbook. Texts of the time referred to it as a presenile atrophy and called it a rarity, careful to distinguish it from the more common *senile* atrophy. Shortly after my conversation with Dr. Manuelides, it was recognized that both forms were the same disease, but the presenile was indeed rare and accompanied by a rapid aging in appearance. This would seem almost certainly to be the dementia that afflicted Ignác Semmelweis.

But such a diagnosis, convincing as it may be, sheds no light on Semmelweis's behavior in the years before his pathology developed. It does not help to explain his long delay in publishing; his sudden flight from Vienna to Pest; the personality changes that others were noting as early as a decade before he could have had organic brain disease; the complete absence of the tact that would certainly have furthered his aims on the obstetric units he directed, and the belligerence with which he approached transactions with skeptical or recalcitrant staff members; or the totality of his identification with the *Lehre*, to the point where he perceived all criticism directed at it to be criticism of him. Nor does it explain the most obvious theme running through his decisions and actions in the years after 1847—the relentless course of his self-destructiveness.

The mythology that has for so long surrounded the life of Ignác Semmelweis made him out to be an iconic figure fighting a lonely battle against the ignorance and arrogance of authority, as though the entire established hierarchy of mid-nineteenth-century medicine had arrayed itself against him. In this construction, only he understood the great truth he was attempting to offer to the world, but it was such a threat to the power structure that he lived in a grand defiance of the controlling figures of his time and of the ordered scientific world they represented. They conspired to bring him down, first driving him from Vienna and then from sanity. His death was the death of a martyred hero, a victim of the very disease he had been fighting with such unalloyed courage and selfless beneficence throughout his career. It provided the narrative of his life's journey with the final dramatic moment that would make of it an epic saga.

The truth is that Semmelweis's battle was lonely because he chose to make it so. His first supporters were among the emerging leaders of Vienna medicine, and they believed in him. But in deserting them he deserted the most powerful armament of his own cause, influential allies. His other allies were few; he did little to encourage colleagues to join his cause, unless loud declamations against their obstinacy and incompetence can be considered appropriate methods of persuasion. The more misunderstanding and opposition he faced, the more defensive he became, in a blustering bombast of self-righteousness that alienated many who might otherwise have looked more willingly at the evidence he finally threw in their faces with the publication of the *Etiology* and the open letters. He was, in the common parlance, his own worst enemy.

Perhaps it is best to leave it at that. After an outline of the facts, it is not necessary to go further, as one must if the cause of the self-destructive behavior is to be sought. But the clues left dangling virtually beg to be examined. Having come this far, I have found them difficult to resist.

Equally difficult to resist is the conclusion that Ignác Semmelweis, throughout his medical career, considered himself an outsider. Of course, Vienna in general and the Allgemeine Krankenhaus in particular had long been a haven for foreigners who came to the center of the empire and its medical Mecca from territories near and far, became assimilated personally and professionally, and went on to significant accomplishments. Among them were Rokitansky and Skoda from Bohemia and Kolletschka and Hebra from Moravia. But whatever accents they may have brought with them were only that—accents. None of the four spoke a dialect of German so distinctive that its constructions and sound would always seem awkward to everyone who heard it outside of its original provenance. And Semmelweis had another language problem, even in his native land—he had never fully mastered Hungarian, in a country of minorities where the Magyars were the national ideal. Writing in 1876, Theodor Billroth, in a section of *The Medical Sciences in the German Universities* entitled "Student Types," would carefully distinguish "German, but also Italian, pure Magyar and Slavic elements at Vienna" as being those preferred, with other groups occupying a gradation of approval. The *Hofrath* pointedly noted that "the Hungarian Jews have the worst reputation among the Viennese students themselves." It may be overstating the argument to wonder whether Semmelweis, a Hungarian with a name easily taken for Jewish, bore yet another stigma of which he was only too painfully aware. In any case, living in either Vienna or Pest, he was somewhat of a stranger. And his language was clumsy, whether he was speaking or writing. No wonder he was so reluctant to put pen to paper in defense of his *Lehre*.

The saga of Ignác Semmelweis can be seen as the consummation of a self-fulfilling prophecy. I would submit that he saw himself as a maladroit, graceless foreigner, who came from the wrong place, spoke the wrong dialect, and became an obstetrician only after being rejected for posts that carried more prestige—in short, always the outsider clanging and banging on the gates of an academic Pantheon in which he felt unworthy to dwell. The reality of his immense discovery and of the power of his friends could not overcome his greater sense of unworthiness. As so often happens in psychopathology, that self-concept existed side by side with its opposite: a growing megalomania, a rage, and finally a towering hurricane of grandiosity that swept him to his destruction.

Semmelweis's flight to Pest can thus be viewed as an almost conscious step on the road that he was inexorably traveling toward self-destruction; furthermore, victory and the attainment of a professorship at the hallowed University of Vienna were inconsistent with his unconscious prophecy for himself; he ran back to Hungary because it was safe and it was home; he made himself believe in the fantasy of his rejection because it gave him the rationalization he needed to rush back to that protective cocoon. Who can know what role was played in all of this by the relatively recent deaths first of his mother and then of his father?

By dint of hard work and a touch of genius, Semmelweis made a monumental discovery and was embraced by some of the brightest stars in the rising new order of Austrian medicine. But they moved easily in this glittering intellectual milieu and had become part of a select group of which he could never see himself a member; he felt unqualified to write or to speak in a public forum. Even when victory was at hand, Semmelweis's concept of being a clumsy outsider was too strong. Seizing on an insult that was in reality only a minor setback, he fled the sunlit arena where just a little more debate, a little more time, and some well-planned laboratory studies would surely have brought him recognition as one of Europe's outstanding clinical investigators. His friends and supporters were dumbfounded. But they could not have known that such apparent disloyalty was in fact the desperate act of a pride-injured and deeply insecure man unable to take the obvious next step toward becoming a celebrated professor in Vienna.

Not only did flight to Pest solve no problems, but it created its own. In the wave of Magyarization, Semmelweis wore Hungarian national costume in the lecture halls and clinics of the medical school, but he was still an outsider, no doubt being certain that the Magyars looked on him as a German and the Germans as a second-class Hungarian. In spite of great success with the application of his theories in Pest, all criticism became intolerable, because it was further confirmation of his sense of the impossibility of victory and the inevitability of failure. And finally, in a fit of growing madness that was partly organic and partly the result of his almost conscious self-prophecy, he became Samson Agonistes, blind and raging, and tried to pull down the pillars of resistance to his *Lehre*, hoping to destroy those whom he saw as his sworn enemies, not realizing that it meant his own immolation. When it was over, only Semmelweis was dead. The temple of resistance still stood.

Semmelweis's many biographers have created a mythology that compares the events of his life to those of a tragedy in the manner of Aeschylus, in which the hero is destroyed by malevolent gods—by forces beyond his control. But the resemblance is far greater to the plays of Sophocles, where the hero's fate is governed not by the actions of the gods but by a fundamental fault in his own nature. The drama of Ignác Semmelweis lies in the fact that, just as he discovered his own truth and his own mission, he created and impelled himself toward his own tragic destiny. This is precisely how Sophocles might have written it, with a Greek chorus of dying mothers—a great hero, a great truth, a great mission, and finally a mad flight of passionate arrogance resulting in destruction. The gods who were the professors of obstetrics did not bring it about; the hero brought it on himself.

Afterword

In 1857, Louis Pasteur—writing in a journal read primarily by his fellow chemists—described having found bacteria in the putrefied material brought to him by a local manufacturer of beetroot alcohol. Realizing that these microbes were the cause of the mysterious catastrophe that was spoiling the products of vintners and brewers in the vicinity of the eastern French city of Lille, the thirty-four-year-old professor carried out further experiments and showed that heating the alcohol to an appropriate degree would kill the germs. The publication of these studies marked the first time any connection had been made between bacteria and pathological changes in organic matter.

In the very year of the death of Ignác Semmelweis, the series of papers subsequently written by Pasteur was brought to the attention of a young English surgeon at the Glasgow Royal Infirmary, who immediately began to study the foul-smelling pus draining from the infected material that was causing a 34 percent amputation mortality at his hospital.

Because his father was the amateur scientist who had solved the problem of lens aberration in 1832, Joseph Lister was an expert microscopist. Peering down at the lethal material of wound after wound, he soon came to the conclusion that microbes were infiltrating healthy tissue and turning it into a quagmire of infection. Reasoning—like Semmelweis before him—that destroying the odor would destroy the lethal material, he began to spray the field of his surgery with dilute solutions of carbolic acid and to dress postoperative wounds with bandages soaked in the same material. His mortality rates dropped by almost two-thirds, and they continued to improve as he perfected his technique. In 1867, Lister published a series of five papers in the British medical journal *Lancet,* announcing the discovery of the new technique he called antisepsis. Without knowing it at the time, he was also announcing the introduction of the germ theory of disease.

In jointly discovering that bacteria cause infection and disease, Pasteur and Lister were fulfilling a long-standing prophecy. Over the course of centuries, more than one medical theorist had predicted that there would one day be found a *contagium animatum,* an invisible organism that explains sickness. In 1546, the Veronese physician Girolamo Fracastoro wrote a treatise called *De Contagione*—mentioned in chapter 3—in which he suggested that microorganisms that he called *seminaria* were the culprit. Using primitive microscopes more than a hundred years later, Anton van Leeuwenhoek observed "animalcules" in water, in water-soaked organic material, and finally in scrapings from his back teeth. But until Lister and Pasteur, no one related them to disease.

Lister, who at that time did not know of the work of

Semmelweis, faced the same opposition—and for many of the same reasons—as did his Hungarian predecessor. But they were two very different men. Lister was a serene English Quaker, who knew that further experimental proof, continued clinical evidence, lucid publications, and the gentle persuasion that was his natural habit would one day win the battle. But his peaceful disposition hid a steely will. He permitted neither forceful disagreement, nor denigration, nor even ridicule to hinder his quiet determination to convince the medical world of the correctness of his theory. Acceptance was slow in coming. It took the better part of two decades and much corroborative work, by him and others, before antisepsis and then asepsis were instituted at increasing numbers of hospitals throughout the world. Lagging slowly behind, the United States did not see its first thoroughly aseptic operating room until the opening of the Johns Hopkins Hospital in 1889. But by then the process was complete. Lister, first knighted and later made a baron for the magnitude of his contribution—to this day considered one of the greatest in the long history of medicine—became an international celebrity. He was showered with medals, honorary degrees, and titles from nations across the face of Europe. And through it all, he never lost the modesty, warmth, and patience that had brought his mission to its successful conclusion.

As a result it was no longer possible to deny the validity of the Semmelweis *Lehre*. As though rediscovered, his barely remembered name now became associated with the notion of genius, first in Hungary and later throughout the world. A man who, in reality, had no lasting effect on obstetric practice became lionized as an icon in the salvation of mothers everywhere. The encomia spilled forth in a profuse stream still strongly in evidence today. In 1891, the medical faculty of the institution by then called Budapest University appointed a Semmelweis memorial committee, charged with honoring the man now recognized as a national hero. In that same year, his remains were brought home from their ignored grave in Vienna and placed first in the mausoleum of his wife's family in the Kerepesi cemetery and three years later under a magnificent tombstone of their own, as part of a grand ceremony addressed by professors of obstetrics from Hungary, France, and England. A major celebration was held in 1906, at which a panoply of international orators lauded Semmelweis on the occasion of the unveiling of a memorial statue of the great man. On the centenary of 1847, festive celebrations were held in Budapest, Vienna, Portugal, Czechoslovakia, Spain, Brazil, Mexico, Switzerland, the United States, and elsewhere. The posthumous honors multiplied as though to compensate for the ignominy in which their subject died.

Finally, in 1963, Semmelweis's remains were exhumed once again. Radiological and other studies were done on them, and they were placed in a courtyard wall of the house where he was born, then being restored as the Semmelweis Medical History Museum. The institution where his work was rejected by so many is now the Semmelweis University of Medicine.

Had Ignác Semmelweis so much as once asked the microscopist Joseph Hyrtl to study a drop of pus from one of the dead mothers, he would have found it to be teeming with the same kinds of organisms that Lister later found in his infected wounds. The invisible organic particles would have been shown to be bacteria. The leap of genius that had allowed Semmelweis to reach this astonishing insight was incalculable in its potential implications. But just as incalculable was the power of obstinate blindness that stopped him at that very point, beyond which he seems constitutionally to have been unable to go. It would remain for others to identify the nature of the lethal microorganisms, while he lay moldering in an unvisited grave.

The process of discovery began in 1869, when two Frenchmen—Coze and Feltz—reported having seen *microbes en chainettes,* chains of microbes (or, as they would later come to be called, streptococci), in the lochia of women with puerperal fever. Ten years later, Louis Pasteur found the same organisms not only in lochia but also in the blood of victims of the disease. If there is such a thing as an immortal soul, one can only imagine how Semmelweis's would have been cheered by the image of the great bacteriologist striding to the platform during a medical meeting in Paris, to interrupt a speaker droning on about his own incoherent views of the etiology of puerperal fever. Drawing a blackboard picture of the streptococci he had demonstrated in blood and lochia, Pasteur unknowingly proclaimed the crux of the Semmelweis *Lehre.* "It is the doctor and his staff," he told his audience with appropriate Gallic certainty, "who carry the microbe from a sick woman to a healthy woman."

It would take decades to elucidate the nature of that microbe, during which much work needed to be done in the classification of bacteria and the understanding of their behavior. At first called streptococcus pyogenes, the germs later became more distinctly categorized as streptococcus hemolyticus and are now known as Group A beta hemolytic streptococcus. Though they are indicted as by far the most common cause of puerperal fever—both as it existed in the

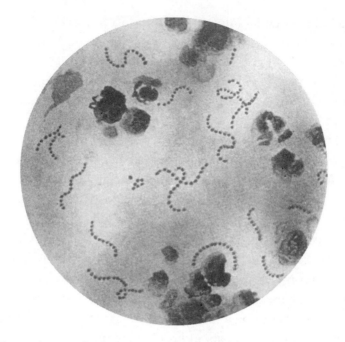

Photomicrograph of streptococcus pyogenes bacteria, the microbes that carry puerperal fever. (Centers for Disease Control and Prevention)

time of Semmelweis and in the rare cases still seen today—other organisms may also be involved, with either the same or somewhat different clinical manifestations.

At a time when bacteria had not yet been recognized as the cause of disease, it was the genius of Ignác Semmelweis to present a theory in which he all but came to that precise conclusion, on the basis of scrupulous observation and the study of medical literature and hospital records. Though he failed to take the next step, he came within a whisper of deducing the germ theory of disease; had his personality been different, he would have discovered it. As it was, his work was neglected and all but forgotten. And so it was the fate of Semmelweis that he was not even a herald of the coming order. To receive his due of honor, he had to be rediscovered. Very likely, there have been Semmelweises before and since his time who have left even smaller traces of their work, or none at all. But at least his name is known and can be hailed, even if by so much too late. The others are lost to history.

I visited the Allgemeine Krankenhaus for the first time in 1985. One of the goals I had set for myself was to find the First and Second Divisions, and walk the actual wards where Semmelweis had made his great discovery. Having spent a few hours with Professor Manfred Skopec of the University of Vienna's Institute for the History of Medicine, I was armed with a map on which locations of historical interest were marked.

According to the map, the building I sought was in the *achten Hof,* the eighth courtyard of the sprawling hospital campus. Since the courtyards bore only poorly marked signs telling which was which, I inadvertently missed the one I was looking for. Needing directions, I stopped a white-coated young woman who proved to be a resident physician surprisingly without English. She scrutinized my map, and, pointing out that I had wandered beyond my destination and into the ninth courtyard, asked me what I was looking for. In the same halting conversational German with which I had asked her for help, I told her that I was trying to find the place where Ignác Semmelweis had worked. My assumption that everyone in this institution would instantly recognize that name proved to be without foundation. I tried to explain to her that he was the nineteenth-century physician who had discovered hand washing to be the key in preventing childbed fever. Again, I was making a baseless assumption: that thus identifying him would assure recognition. And again, I was wrong. With index finger resting on her lips as though in deep concentration, she thought a moment and then all at once looked at me brightly, a perfect idea having obviously just come to her. *"Sie müssen zum achten Hof zurückgehen,"* she instructed, *"und dort einen alten Arzt finden"*—(You must go back to the eighth courtyard, and find there an old doctor). Needless to say, I was dumbfounded. Clearly, American youth was not alone in the absence of its sense of history or even the great passages of time.

I thanked the young doctor, reversed my steps, and reentered the courtyard through which I had just passed. The moment I stepped into it, I knew that I was in the right place. Preoccupied with my map, I had on my first walk-through not noticed the stone steps leading up to a great door that could be no other than the entrance to the staircase to the two divisions. Feeling the excitement mount, I opened the door and climbed to the first floor, in my mind's eye seeing the many heavy-laden young women who had negotiated these stairs in times past.

On reaching the top, I did not find the expected open space or anything recognizable as an entrance to one or the other division. The entire area where they were had been partitioned off into separate offices and rooms, so that only a narrow corridor remained along the outer walls of what had once been the wards. Hearing voices, I traced them to an office near the entrance to what had obviously been the Second Division, in which four or five students were chatting over their books and afternoon coffee. My English being again of no use, I explained my quest in German, to blank stares. After a brief and almost incomprehensible conversation among themselves, one of the students said that he would go for an orderly who seemed to know something about the history of the place. He came back with a pleasant middle-aged man in tow, to whom I explained what I was looking for. "Of course," he replied, to my great relief. "Come with me." I followed him to the former First Division, where, halfway down the corridor and high on the wall—where it was easily missed unless one was deliberately looking upward—stood a plaque proclaiming that Ignác Semmelweis, working in this place, had made his great discovery in 1847. The orderly knew about the plaque, but not about the man. Laden with posthumous honors in his native land, Semmelweis was unknown in the very corridors he had walked here at the Allgemeine Krankenhaus.

My entire pilgrimage had been an exercise in frustration. I asked the orderly to take some photos of me under the plaque, stuffed the map into my pocket, and left.

Bibliographical Notes

Unless one reads both Hungarian and German fluently, some of the source material concerning the life and death of Ignác Semmelweis is inaccessible. But there have been very good translations of a good portion of it, and there exists a strong secondary literature, much of which has appeared within the past two or perhaps three decades.

Though outdated and scattered with information later shown to be questionable, the exhaustive biography published by William Sinclair in 1909 remains an excellent introduction to Semmelweis studies. Writing long before the current era of hero debunking, Sinclair produced a book that is sometimes worshipful, but always respectful of the facts as they were known at that time. And that time, in fact, was the height of the Semmelweis idolatry, approximately the period of many memorials and celebrations in his honor, when newly discovered and usually commendatory information was everywhere in the air. There is something to be said for

being introduced to a historical figure by an author convinced of his subject's greatness, and that is certainly the situation in Sinclair's detailed and very readable volume.

Sinclair's contribution strikes a note quite different from
the other laudatory one published in English translation by
György Gortvay and Imre Zoltán in 1968. They can best be
described as determined Hungarian apologists for Semmelweis, which explains some of the glaring hagiographic weak-
ness of their book. They tend to squeeze facts into poorly
fitting pigeonholes, producing not only slanted images but
also interpretations that make their subject and the Hungarian institutions with which he was affiliated look far better
than they were. Despite the great Hungarian accomplishments in poetry, literature, music, mathematics, physics, and
chemistry, Semmelweis is to this day the only product of that
fascinatingly productive culture to be of major consequence
in the history of medicine, and Gortvay and Zoltán make the
most of him. Their book must be read with the enthusiasm
turned down a notch. Its chief virtue is that it presents information from uniquely Hungarian sources, which sometimes
has an immediacy that brings settings and people to life on
the page, even in the daunting face of Zoltán's somewhat
stilted translation.

I have used the first English translation of the *Etiology,* produced in 1941 by Frank B. Murphy, professor of obstetrics at
Creighton University. As Murphy wrote in his introduction,
"The style is wordy and repetitious; the argument flows back
and forth without progressing to any logical point; the author
is egotistical and bellicose. . . . If Semmelweis had only spent
more time in clearly stating his views and less in argument,
his book would be twice as good and half as long!" The quo-

tations from the open letters are my own, done in conjunction with Ferenc Gyorgyey and published in 1981 by the
Classics of Medicine Library. To my knowledge, they remain
the only source for an English version of these writings.

I have referred to K. Codell Carter as a leading academic
Semmelweis scholar, and so he is. This in spite of our dis-
agreement about certain interpretive issues. Written with his
wife, Barbara, as co-author, his 1994 book, *Childbed Fever: A
Scientific Biography of Ignaz Semmelweis,* published by the
Greenwood Press, is a somewhat popularized collection of
previously published essays that provide a sound biography
and more, especially for the general reader. In addition,
Carter, a professor of philosophy at Brigham Young
University, has done the only English translation—other than
Murphy's—of the *Etiology,* published in 1983 and preceded
by a useful 58-page biography.

For readers who would like to learn more about the story
of puerperal fever, no source is better than the writings of
Irvine Loudon, and most specifically his recent *The Tragedy
of Childbed Fever,* which is the only truly comprehensive
study ever produced about this fascinating disease and its
complex history. Loudon's earlier publication, *Childbed
Fever: A Documentary History,* reproduces twenty-two primary sources, in essays and extracts from the work of authors
from 1772 to 1968.

For many years, I have benefited from reading in Erna
Lesky's monumental history of the golden years of Austrian
medicine, *The Vienna Medical School of the 19th Century,*
published in English by the Johns Hopkins University Press
in 1976. Not only is it the most authoritative and detailed
source of information about that glorious place and time,

but it excels in conveying the atmosphere of an era. None of its passages are more instructive than those in which Professor Lesky describes the contributions, personalities, and interactions of such individuals as Rokitansky, Skoda, Hebra, Klein, Braun, and Semmelweis himself.

Not surprisingly, the Semmelweis story has captured the imagination of several popular writers, including Louis Ferdinand Céline, who wrote his somewhat confused doctoral thesis on the subject. Frank Slaughter's *The Immortal Magyar* is a nontechnical biography for the general reader, and Morton Thompson's 1949 best-seller, *The Cry and the Covenant,* is a fictionalized account of the major events.

My fascination with Ignác Semmelweis began more than twenty-five years ago. The earliest results of immersing myself in his life were a lecture in the Yale History of Surgery series in February 1978 and then the publication, in a 1979 issue of the *Journal of the History of Medicine and Allied Sciences,* of my essay "The Enigma of Semmelweis—an Interpretation." This paper, as I discovered when I visited historians and the Semmelweis Museum in Budapest a few years later, disturbed a few worshipers at the shrine of the Hungarian national medical hero. Objections were raised by some about my description of Semmelweis's character, particularly its self-destructive aspects. Also, no one appeared to be completely happy with the diagnosis of Alzheimer's, though it seemed virtually self-evident at the time and still does. I still wonder why my hosts should have thought syphilis more palatable. And while my thesis about the asylum beating had by then begun to gain some currency, it was hardly a universally accepted opinion. But I was convinced that the evidence bore me out on all the issues I had raised, and time has only strengthened that viewpoint. So certain had I become, after being confronted with the overly vigorous Hungarian denials, that I republished the essay in my 1988 book on the history of medicine for the general reader, *Doctors: The Biography of Medicine.*

Not only that, but I have leaned heavily on that 1979 paper as this book was being written. My basic arguments are the same as they were then, and I feel more strongly than ever that they are valid. Small bits of information that have become available since that time have only fortified my convictions and increased the number of like-minded people. I have incorporated a few paragraphs of the original essay and perhaps that many additional sentences into the present book. They help tell the story as I have increasingly come to see it.